# Life Beyond Earth
## The Search for Habitable Worlds in the Universe

With current missions to Mars and the Earth-like moon Titan, and many more missions planned, humankind stands on the verge of exciting progress and possible major discoveries in our quest for life in space.

What is life and where can it exist? What searches are being made to identify conditions for life on other worlds? If extraterrestrial inhabited worlds are found, how can we explore them? Could humans survive beyond the Earth?

In this book, two leading astrophysicists provide an engaging account of where we stand in our quest for habitable environments, in the Solar System and beyond. Starting from basic concepts, the narrative builds scientifically, including more in-depth material as boxed additions to the main text. The authors recount fascinating recent discoveries, from space missions and observations using ground-based telescopes, of possible life-related artefacts in Martian meteorites, of extrasolar planets, and of subsurface oceans on Europa, Titan and Enceladus. They also provide a forward look to exciting future missions, including the return to Venus, Mars and the Moon; further explorations of Pluto and Jupiter's icy moons; and placing giant planet-seeking telescopes in orbit beyond Jupiter, showing how we approach the question of finding out whether the life that teems on our own planet is unique.

This is an exciting, informative read for anyone interested in the search for habitable and inhabited planets, and makes an excellent primer for students keen to learn about astrobiology, habitability, planetary science and astronomy.

ATHENA COUSTENIS is Director of Research at the French National Research Center (CNRS), and an astrophysicist at the Laboratoire d'Etudes Spatiales et d'Instrumentation en Astrophysique (LESIA) of Paris Observatory. She is Co-investigator of three of the instruments

(CIRS, HASI, DISR) aboard the Cassini–Huygens mission. Her expertise in space missions has allowed her to chair or to contribute to several advisory groups within ESA and NASA. Dr Coustenis is currently President of the International Association of Meteorology and Atmospheric Sciences, as well as Secretary of the Division for Planetary Sciences Committee. She is a member of several editorial boards and has received several NASA and ESA achievement awards.

THÉRÈSE ENCRENAZ is Emeritus Director of Research at CNRS, and an astrophysicist at LESIA, Paris Observatory. She has been involved in many planetary space missions, and has been a Mission Scientist of the European ISO (Infrared Space Observatory) mission. She has chaired the Science Advisory Committee of CNES for the exploration of the Universe. She is currently a member of the E-ELT Project Science Team. Dr Encrenaz is the author of more than 250 refereed articles, a few lecture books and a dozen popular books. She has received several awards including the Silver Medal of CNRS and the David Bates Medal of the European Geophysical Union.

# Life Beyond Earth

## The Search for Habitable Worlds in the Universe

ATHENA COUSTENIS

and

THÉRÈSE ENCRENAZ

*Laboratoire d'Etudes Spatiales et d'Instrumentation en Astrophysique, Paris Observatory, France*

CAMBRIDGE
UNIVERSITY PRESS

# CAMBRIDGE
## UNIVERSITY PRESS

University Printing House, Cambridge CB2 8BS, United Kingdom

Published in the United States of America by Cambridge University Press, New York

Cambridge University Press is part of the University of Cambridge.

It furthers the University's mission by disseminating knowledge in the pursuit of education, learning, and research at the highest international levels of excellence.

www.cambridge.org
Information on this title: www.cambridge.org/9781107026179

© Athena Coustenis and Thérèse Encrenaz 2013

First published 2013

Printing in the United Kingdom by TJ International Ltd. Padstow Cornwall

*A catalogue record for this publication is available from the British Library*

ISBN 978-1-107-02617-9 Hardback

Cambridge University Press has no responsibility for the persistence or accuracy of URLs for external or third-party internet websites referred to in this publication, and does not guarantee that any content on such websites is, or will remain, accurate or appropriate.

# Contents

Colour plates section can be found between pages 150 and 151.

# Preface

Life in space, whether strange beings on distant worlds, or an expansion of our own species into the Solar System and beyond, is a very exciting idea. Humankind may currently stand on the verge of major discoveries and exciting progress in both areas. The discoveries of possibly life-related artefacts in a Martian meteorite, in a subsurface ocean on Europa, Titan or Enceladus, and in the atmospheres of extrasolar planets, for example, show how close we are to finding out at last whether the life that teems on our own planet is unique. Some increasingly sophisticated space missions are currently under way, such as Cassini, which has been exploring the Saturnian system and Titan, the Earth-like moon, since 2004; others are in preparation, such as the Mars Sample Return and the Jupiter Icy Moons Explorer missions. Plans to return to Venus, Mars, the Moon and Titan, to orbit Europa and to place giant planet-seeking telescopes in space are thus on the table. These and other advances promise rapid progress in the coming years.

This is a book that deals with possible habitats in our Solar System and beyond. We will define which places might be harbouring past, present or future life, or can be considered as 'habitable' in the sense that human life could survive, adapt or continue to evolve therein. The book will include a necessarily brief but pertinent definition of life as we know it on Earth and review it as a phenomenon, considering its origins, properties and potential; we will combine a discussion of present knowledge with informed speculation, bounded by scientific realism but using non-technical language. We will briefly review the origin of life in the Universe, the reasons for thinking it may be unique and reasons, in contrast, for believing it could be commonplace. We will also offer some thoughts on its destiny and on scientific discoveries yet to be made in areas we can barely apprehend at present. The main goal is to update the reader on the current situation in our Solar System and beyond, in terms of exploration for traces of past or present life and of the existence of conditions for habitable worlds. We also aim to provide and provoke thoughts about our distant horizons in this respect.

The format of the book is such as to address a large audience (lay persons, students and others). The purpose is not to give an exhaustive description from the biological, geological or philosophical point of view, but rather to excite the imagination of the reader, by including up-to-date illustrations and clear, relevant and accurate text that only astrophysicists can provide on recent discoveries and future projects. As astronomers, we will offer a personal, inside view of space exploration, using our own knowledge and interests to describe the most interesting places outside Earth, as well as the vanguard techniques that we use to investigate them. We would like to thank here all of our colleagues (experts in various fields of astronomy) who assisted us with information, discussions and re-reading, and the artists who gracefully provided us with figures and photographs.

# I    Introduction

The search for life in the Universe, from theoretical concept to actual exploration, has never ceased to interest and amaze humanity. After the first ideas had arisen on cosmology (the structure of the cosmos) and cosmogony (its creation), early civilizations and philosophers turned their minds towards living beings and how they came to be. Once some basic principles had been set – for instance in the biblical book of Genesis or in Hesiodos' *Theogony*, which both basically define the creation of Earth and Heavens from nothing (or Chaos) – the first 'scientific' minds set to work all over the world, and new ideas were sparked in Egypt, in the Indies, in the Americas, in China and in Europe. Thus, in Greece for instance, Aristarchus conceived the idea of the heliocentric Solar System; Eratosthenes proved that the Earth was spherical and determined the distance to the Moon; and Anaximander had a structure worked out for the whole Universe.

Some of the early thinkers had already advocated a Universe consisting of 'many worlds'. Thales, from Miletus, and his students in the seventh century BCE argued for a Universe full of other planets, teeming with extraterrestrial life. They also proposed the idea with which we are all familiar today (through Drake's equation, Carl Sagan's musings, and the contributions of many other scientists): that a Universe so full of stars must also have a large number of populated worlds. This proposal was defended by Epicurus and other Greek atomists who countered the geocentric models put forward later by Aristotle. In the cosmogony developed by Plato's famous disciple, the mythological separation of Earth from the Heavens was put into more modern words and widely promoted, as was his geocentric perception of the cosmos and the limited and well-defined sphere of stars in which matter and space were confined and interconnected. Aristotle's philosophical attempt at modern physics took strong roots, caused the ancient open-minded theories to be forgotten and hindered scientific progress in this domain for quite a long period. The Copernican revolution in the sixteenth century gave a boost to the concept of life's emergence and possible existence elsewhere in the Universe, because Earth was no longer the privileged and unique place where this could occur.

In 1862, the French scientist Camille Flammarion published *La pluralité des mondes habités* ('On the plurality of inhabited worlds'), in which he discussed the conditions of habitability and the possible presence of life on such habitable planets of our Solar System. The public loved the book, but Urbain Le Verrier (then Director of the Paris Observatory) and many of his colleagues utterly rejected Flammarion's arguments, and Flammarion was consequently fired from the Observatory. Open-mindedness was not always accepted at that time, but – thankfully – we have come a long way since then.

## 1.1   THE QUEST FOR LIFE

One hundred and fifty years later, in the era of planetary exploration, with space missions and large telescopes at our disposal, the quest for life remains just as important to humanity. The discovery in the past 20 years of planets around other stars (exoplanets) has made a difference in our perception of the possibility that other worlds might harbour life or the conditions necessary for life to emerge and survive.

But, as the quest for life-supporting conditions in our Solar System moves onwards, with ever more powerful means, it is essential to know exactly what we are searching for. Such type of investigations have in the past been driven by geocentric considerations. Robotic exploration surveys in the Solar System, and several astronomical surveys from the ground and space targeting 'exoplanets' (planets beyond our Solar System), have been designed to retrieve information on present or past signatures of life or biotic-related elements, but up to now these have always been based on life as we know it on Earth – terrestrial life (Figure 1.1). The modern definition of the main features

FIGURE 1.1   The Earth from space. The presence of water is detectable remotely from the dark blue colour of the oceans, the water vapour clouds and the northern polar cap. For colour version, see plates section. (Image courtesy of NASA/GSFC/Suomi NPP.)

of life, as found on our planet, includes the presence of liquid water, energy sources, a stable environment and nutrients. We discuss these elements and their relation to life in the next chapter, but there is no doubt that one of the main ingredients, identified as such early on, is liquid water.

It is not surprising, then, that up to now robotic space exploration has been directed to places in the Solar System where liquid water is possible (and more specifically in exposed locations on the surfaces of planetary bodies), and that it emphasizes searches for formations that resemble terrestrial organisms. Even though this approach is understandable, given how little we know about life's origin, biology theories and experiments are now pointing to the fact that living organisms elsewhere may well be quite different from terrestrial life. These new scientific studies seem to indicate that if life appeared elsewhere in our Solar System and beyond, we may need to broaden our exploration designs in order to accommodate the possibility of non-terrestrial-like signatures of current or past biotas. Such organisms may not be using liquid water as a solvent; and simply from a pure physical and not biological point of view, the water may not be located on the surfaces of the planetary bodies but elsewhere. So we may be currently neglecting the exploration of planetary environments which may well be potential hosts for life.

The authors of this book are not biologists, but planetary astrophysicists. Although we do discuss the aspects above, we are mainly concerned with offering the viewpoint of astronomers and observers of our Solar System on where different environments can be found and explored, often in comparison with our own planet, but also as a framework for establishing the conditions that led to the creation of our Solar System. Here we will not deal so much with the search for life itself, although we will necessarily touch on the subject, but rather with the habitability conditions that we can expect to find in the Solar System and beyond, in other stellar systems.

## 1.2   THE FORMATION OF PLANETS

Before asking whether life might have appeared in other planets of the Solar System, we need to understand how planets formed. Let us start by considering the most recent theories on the formation and evolution of our neighbourhood. In this, we are helped by the observation of other nearby young stars. Over the past decades, observations of star-forming regions have revealed that more than 50 per cent of young stars are surrounded by a protoplanetary accretion disk (Figure 1.2). It is generally accepted that planets outside our Solar System, now discovered in their hundreds, formed within these disks.

A similar story is told by the Solar System itself. A few basic observations, made as early as the eighteenth century by Immanuel Kant (1724–1804) and later by Pierre-Simon de Laplace (1749–1827),

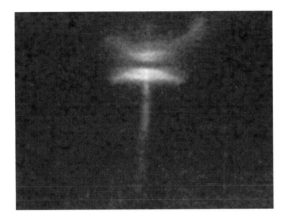

FIGURE 1.2 A protoplanetary disk, HH-30 in Taurus, about 450 light years away, observed with the Hubble Space Telescope. The disk emits a stellar jet, aligned with the rotational axis of the disk which appears in black as an absorption in front of the stellar light. For colour version, see plate section. (Image courtesy of STSci/ESA/NASA.)

suggest that Solar System planets formed within a protosolar disk, resulting from the collapse of a protosolar cloud in fast rotation around an axis perpendicular to the disk plane. Indeed, the orbits of the planets show a few remarkable properties: they are almost coplanar, circular and concentric around the Sun, and they all rotate in the same direction, as does the Sun. It is generally accepted that Solar System planets formed by accretion of solid particles, through mutual interactions, collisions and gravity (Figure 1.3). While the first steps of the accretion process can be reproduced in a satisfactory way by models of interactions between gas and dust in the protosolar disk, it is more difficult, at present, to understand how planets crossed the metre-size barrier. For larger sizes, the planet's gravity becomes sufficient to accrete the surrounding material, explaining the runaway growth of the biggest embryos.

Another remarkable property of the Solar System planets is their clear division into two classes: the terrestrial planets, relatively close to the Sun, and the more distant giant planets (Figure 1.4). The terrestrial planets – Mercury, Venus, the Earth and Mars – are characterized

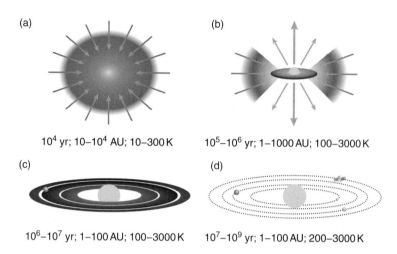

(a)

$10^4$ yr; $10$–$10^4$ AU; $10$–$300$ K

(b)

$10^5$–$10^6$ yr; $1$–$1000$ AU; $100$–$3000$ K

(c)

$10^6$–$10^7$ yr; $1$–$100$ AU; $100$–$3000$ K

(d)

$10^7$–$10^9$ yr; $1$–$100$ AU; $200$–$3000$ K

FIGURE 1.3 A schematic view of the formation of planets in the protosolar nebula. (a) A molecular cloud contracts and rotates at an increasing speed. (b) The cloud collapses into a disk perpendicular to the rotation axis of the cloud; the central matter accretes to form the proto-Sun. (c) Planetesimals form within the disk from the accretion of solid particles. (d) After the dissipation of the gaseous disk, a few planets and small bodies are left. (Image adapted from NASA/James Webb Space Telescope.)

by relatively small sizes, high densities (from 3.9 to 5.5 g/cm$^3$), a solid surface and very few satellites. In contrast, the giant planets, Jupiter, Saturn, Uranus and Neptune, have large radii (from 4 to 11 Earth radii), a low density (from 0.7 to 1.6 g/cm$^3$), a thick hydrogen-dominated atmosphere and a large number of satellites, with many of these being 'regular' (i.e. in the equatorial plane of the planet). The origin of this dichotomy has to be found in the way these objects formed and evolved.

## 1.2.1 Formation of the Solar System

We have first to remember that the protosolar disk composition must have reflected cosmic abundances: hydrogen was by far the most abundant element. It was formed, together with deuterium, helium, lithium and beryllium, at the time of the Big Bang, by primordial nucleosynthesis. The heavier elements – in particular carbon, oxygen

FIGURE 1.4 A schematic view of the Solar System (not to scale), with the terrestrial planets close to the Sun and the giant planets at further distances, beyond the main asteroid belt. A comet is shown near the orbits of Jupiter and Saturn. For colour version, see plates section. (Image courtesy of NASA/JPL.)

and nitrogen – were formed by nuclear reaction within the stars, in abundances which, to first order, decreased as their atomic number increased (Figure 1.5). The heavier elements were thus the less abundant ones.

Within the protosolar disk, the temperature decreased as the distance from the Sun increased. Two cases can be considered: in the vicinity of the Sun, where the temperature was above about 200 K, the only solid material was silicates, oxides and metals. Because these elements are intrinsically rare, the solid mass available for planetary embryos was limited. Only planets of the size of the Earth or less could form, with a relatively high density, typical of silicates or metallic compounds. Their atmosphere represented only a very small fraction of their mass; it was probably accreted in a second step, partly by outgassing (the release of gas from the interior of planetary bodies,

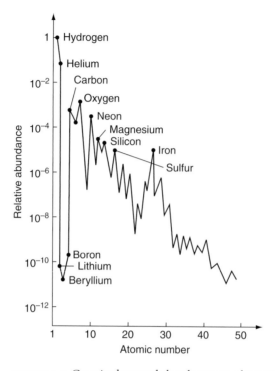

FIGURE 1.5 Cosmic elemental abundances as a function of the atomic mass number of the elements. The abundances are normalized to hydrogen. (Adapted from D. Darling, www.daviddarling.info.)

gas that was perhaps trapped in the primordial material during its formation) and partly from material brought in by meteoritic or cometary impacts.

In contrast, at distances more than five times the mean distance from Earth to the Sun (known as an 'astronomical unit' or AU; the AU is the semi-major axis of the Earth's orbit, i.e. 149.6 million km) where the temperature was lower than about 200 K, the most abundant elements apart from hydrogen and helium (i.e. carbon, oxygen and nitrogen) were in the form of ices ($H_2O$, $CO_2$, $NH_3$, $CH_4$...). These elements were abundant enough to form big solid nuclei. Calculations show that when the mass of such nuclei becomes larger than 10–15 Earth masses, the gravity field is sufficient to pull in or accrete the surrounding nebula, mostly composed of gaseous hydrogen and helium. This is how the giant planets were formed, with very large sizes and relatively low densities. After the collapse phase of the surrounding nebula, regular satellites were formed in the equatorial plane of the planets, and others were captured by the large gravity field of the big planetesimals (small objects formed from dust, rock and other materials, thought to have orbited the Sun at the beginning of Solar System formation, and serving as the building blocks of the planets and satellites by gravitational aggregation), which explains the large number of outer satellites orbiting the giant planets.

### 1.2.2   Migration in the Solar System

Other open questions remain. One of them has to do with the two different classes of giant planets: Jupiter and Saturn on the one hand, and Uranus and Neptune on the other. With masses of 318 and 95 times that of the Earth, Jupiter and Saturn are mostly made up of their protosolar gaseous components, hydrogen and helium (and are thus known as 'gas giants'). In contrast, Uranus and Neptune, located at further distances, with masses of 15 and 17 terrestrial masses respectively, mostly consist of their icy core ('ice giants'). What causes this difference? It has been argued that Uranus and Neptune, located in a less dense region of the disk, needed more time – possibly 10 million years – for their icy core to reach the critical limit of 10 terrestrial masses. After 10 million years, the protosolar disk may have dissipated as a result of the increasing activity of the young Sun (known as the T-Tauri phase), and little gas would have then been available for the accretion phase of Uranus and Neptune. But the true story may have been even more complex.

Indeed, although it was originally believed that planetary orbits had been stable all through the Solar System's history, the consensus today (following the early work of Henri Poincaré and more recent studies) is that unstable and chaotic situations have occurred in the past and may occur in the future. On such occasions, small gravitational perturbations may have induced very strong effects on the motions of planetary bodies. Dynamical simulations support this hypothesis and suggest that very early on, at the time of the planetary formation phase or shortly afterwards, gravitational interaction with the gas in the disk may have led to significant changes in the orbital parameters of the planets. Following dynamical numerical simulations performed, in particular, at the Nice Observatory in France (the so-called 'Nice model'), it seems that all giant planets may have migrated somewhat in the early phases of their history (e.g. Walsh and Morbidelli, 2011).

Estimations of the radial migration and mass growth imposed on the giant planets through simulations seem to indicate that a fully formed Jupiter started at 3.5 AU, a location presumably favourable for giant planet formation owing to the presence of the so-called 'snow line', the distance beyond which water can condense. Saturn's core, with a mass of 30 Earth masses (an Earth mass is often given the symbol $M_\oplus$), initially lies at 4.5 AU. It grows to 60 $M_\oplus$ as Jupiter migrates inwards, over $10^5$ years. Inward migration is stopped for planetary cores smaller than 50–60 $M_\oplus$, when they attain an equilibrium radius in the disk (where migration forces cancel out mutually), so that Saturn's core remains at 4.5 AU during this phase. Similarly, the cores of Uranus and Neptune begin at 6 and 8 AU and grow from a few $M_\oplus$ without migrating. Once Saturn reaches 60 $M_\oplus$ its inward migration begins, and is much faster than that of the fully grown Jupiter. Then, on catching Jupiter, Saturn is initially trapped in the 3:2 resonance (which means that the revolution periods are in the ratio 3/2, Jupiter revolving precisely three times around the Sun while Saturn manages only two). In the model we are considering, by the time Saturn enters into resonance, Jupiter has migrated inwards as close as 1.5 AU from the Sun, thus stopping the growth of Mars; this scenario therefore explains why the mass of Mars is significantly smaller than expected at this distance from the Sun. The capture in resonance spectacularly changes the migration pattern.

Jupiter and Saturn start migrating outwards together, still staying in their 3:2 resonance. In passing, they capture Uranus and Neptune in resonance (3:2 for Saturn:Uranus and 4:3 for Uranus: Neptune), and these planets are then pushed outwards as well. This outward migration continues as long as there is gas in the disk. At the disappearance of the gas, therefore, the planets are left on mutual resonant orbits, very close to each other and almost circular in shape.

This situation later evolves, under the effect of gravitational perturbations by the remaining matter located outside Neptune's orbit (the planetesimals in the proto-Kuiper Belt, which was probably much more massive than the Kuiper Belt today, as we observe it on debris disks around young nearby stars). As a result of these perturbations, eventually the planets are extracted from their mutual resonances. When this happens, the orbits of the giant planets become unstable. The planets start to have mutual close encounters, and their orbits repel each other, increasing their mutual distances and eventually reaching their current location. Consequently, Neptune, propelled outwards, penetrates into the proto-Kuiper belt. Most of the small bodies of the Kuiper Belt are dispersed outside the Solar System while others are sent off on very eccentric orbits. This scenario explains the current low mass of the Kuiper Belt, as compared with similar debris disks around other stars, and its peculiar structure with different populations, including the dispersed disk and objects in resonance with Neptune. This scenario also explains the 'Late Heavy Bombardment', a spike in the bombardment rate that occurred approximately 4.0 – 3.8 Ga (billion years ago), whose manifestation is observed on the surfaces of all bare Solar System bodies: the Moon, Mars, Mercury, asteroids and giant planet satellites. Indeed, numerical simulations show that it is responsible for the dispersal of the proto-Kuiper belt as well as the removal of many asteroids from the main asteroid belt, resulted in a high collision rate all over the Solar System.

The orbital history of terrestrial planets was also strongly influenced by the drastic change in the orbits of the giant planets. Again, dynamical models show that the inclinations of the rotation axes of some of the planets became unstable (the inclination of the orbit of a planet is defined as the angle between the plane of the orbit of the planet and the ecliptic). Consequently, they started to change rapidly (typically over a few million years), with important implications for their climatic evolution. This was definitely the case for Mars, where traces of fossil glaciers have been identified near the equator, dating from a period where the inclination of the rotation axis was about 45 degrees. The Earth escaped this fate thanks to the presence of its large satellite, the Moon, which stabilized the planet's inclination and thus the evolution of its climate. The orbits of the terrestrial planets also became chaotic. The chaotic effects prevent us from predicting exactly the future evolution of the terrestrial planets' orbits over the lifetime of the Solar System, although dynamical models suggest a very low probability of future collisions among the terrestrial planets.

### 1.2.3  Elemental and isotopic abundances as insights to the formation of the Solar System

In order to understand the history of Solar System bodies, we have to be able to estimate their ages. Some information is provided by the cratering rates at the surfaces of bare bodies: the ancient terrains are more heavily cratered (this method supplies evidence for the Late Heavy Bombardment event previously mentioned). Another method consists in measuring elemental and isotopic abundance ratios, either by remote sensing spectroscopy or, in the case of lunar and meteoritic samples, by *in situ* or laboratory experiments. Dating of these samples is obtained from the abundance measurements of long-lived radioactive isotopes with respect to their stable isotopes. An example is the $(^{87}Ru, {}^{87}Sr)$ system: $^{87}Ru$ disintegrates into $^{87}Sr$ with a half-life of $4.7 \times 10^{10}$ years, a period comparable with the lifetime of the Solar System. The stable isotope $^{86}Sr$ is used as a reference. Isotope ratios of stable elements can also be used to retrieve information on formation and evolution processes in Solar System bodies. As an example, heavier isotopes (for instance, D versus H, or $^{13}C$ versus $^{12}C$) have a tendency to be enriched at low temperature, owing to reactions between ions and molecules occurring at those low temperatures; such an effect is measured, in the case of D/H, in the interstellar medium as well as in laboratory experiments.

Measuring isotopic abundance ratios in small molecules (such as $^{13}C/^{12}C$ in $CH_4$ or $CO_2$, $^{18}O/^{16}O$ in CO or $CO_2$, $^{15}N/^{14}N$ in $NH_3$ or $N_2$, or D/H in $H_2O$ or $CH_4$) can thus place constraints on the formation and evolution processes of the Solar System objects in which they are measured (in particular in planetary or satellite atmospheres and comets). We discuss below in more detail the case of the deuterium to hydrogen (D/H) ratio, which happens to be a powerful tool for constraining the early history of the Solar System. As mentioned previously, deuterium was formed in the Big Bang; in contrast to helium, it is continuously destroyed in the stars, so its mean abundance, over the age of the Universe, can only decrease. The primordial value of D/H is estimated to be about $3\text{--}4 \times 10^{-5}$. Its value in the primordial (or protosolar) nebula (the region of gas in our space neighbourhood which underwent gravitational collapse, and which eventually formed our Sun and the Solar System), measured in the solar wind, is $2.1 \times 10^{-5}$. Its present value in the interstellar medium is about $1.6 \times 10^{-5}$. It is interesting to compare the D/H ratio (measured from $CH_3D/CH_4$) in the giant planets with the protosolar nebula. The ratio in Jupiter and Saturn is found to be close to $2 \times 10^{-5}$, which is consistent with the fact that both giant planets are mostly made of protosolar gas. In contrast, D/H in Uranus and Neptune is higher, with a value of about $5 \times 10^{-5}$; this enrichment illustrates the fact that the ice giants, as indicated by their names, are mostly made up of an icy core.

A comparable story is told by the D/H ratio as measured in water, i.e. from the $HDO/H_2O$ ratio. We first note that the D/H measured in the terrestrial oceans (known as the 'Vienna Standard Mean Ocean Water' or VSMOW value) is $1.56 \times 10^{-4}$, about seven times higher than the protosolar value. What is the explanation for such an excess? The current theory is that the Earth's atmosphere was not outgassed from the planet, but captured from meteoritic or cometary impacts; these impactors, coming from colder regions of the outer Solar System, were enriched in deuterium. Indeed, the D/H ratio measured in some comets shows values as high as $3 \times 10^{-4}$, twice the terrestrial value. This is higher than the value observed in asteroids closer to the Sun, implying that comets were formed in a very cold environment. The question of the origin of water on Earth is discussed in more detail in Section 3.4.

Other very different processes may lead to strong isotopic fractionation. As an example, the D/H ratio measured in water from HDO/$H_2O$ in the atmospheres of Venus and Mars is strongly enriched with respect to the terrestrial value (VSMOW) – by a factor of more than 100 in the case of Venus, and to a lesser extent (a factor of 5) in the case of Mars. The reason is a differential escape rate of water over the history of the two planets: HDO, heavier than $H_2O$, escapes at a slightly slower rate. The high values of D/H on Venus and Mars thus demonstrate that water was much more abundant in the past than today (see Sections 3.2 and 3.3).

A third example of isotopic fractionation is given by the effect of photosynthetic reactions on carbon on Earth, which tend to enrich $^{12}C$ with respect to $^{13}C$; as a result, the $^{13}C/^{12}C$ ratio in living organisms is slightly higher than in minerals (see Subsection 2.1.6).

## I.3 LOOKING FOR WATER

There is a limit between the formation regions of terrestrial and giant planets: it is the region where light molecules (with a few atoms) start to condense. Among these small molecules (water, ammonia, carbon dioxide, hydrogen sulfide, methane...), water has a special property: it is the first to condense as the temperature decreases and can then be found as ice at smaller heliocentric distances than other molecules (Figure 1.6). Water condensation actually defines the 'snow line' which separates, in the protosolar disk as in any protoplanetary disk, the formation regions of the two classes of planets. We must keep in mind that this is a first-order classification: as described above, migration may affect this scenario and render the history of planets much more complex.

We need to go beyond our natural tendency to think that Earth is in the centre of the Universe and instead keep an open mind and embrace novel ideas of where life is possible and even what other forms it might take. Chemical experiments already caution us against 'terracentricity' by demonstrating that some chemical reactions involving non-carbon compounds and using solvents other than

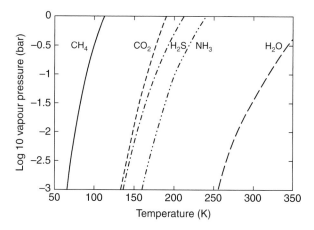

FIGURE 1.6 Saturation curves of a few simple molecules. Among the light molecules, water is by far the first molecule to condense as temperature decreases. (Adapted from Encrenaz, 2008.)

water, or simply non-redox ones exist that could lead to life taking a different form.

Scientists and lay people alike have become interested in well-founded perspectives on the possibility of life in environments in the Solar System and beyond that are very different from the ones that support life on Earth. Some are eager to explore possibilities for life supported by weird chemistry in exotic solvents and exploiting exotic metabolisms. The public has become aware of some of those ideas through science fiction and non-fiction, such as Peter Ward's *Life as We Do Not Know It* (Ward, 2007). As an example, it is possible to conceive of alien life in environments other than the surface of a rocky planet, which would feed on completely different forms of nutrients and get its energy from supersaturated solutions. But most people are essentially wondering whether we are alone in the Universe, and in our modern times we are beginning to move from philosophical to scientific arguments on the subject.

The quest is long and relies on the process of trial and error: the canali on Mars and the false planet around Barnard's star, for instance, have given us a word of caution. But our persistence is rewarded as we break ground in several scientific domains such as astronomy, physics and the new science of exobiology. The revolutionary discovery of planets around other stars since 1995 (over 900 such exoplanets are known today) has broadened our minds and turned a science fiction dream into reality. Protoplanetary disks around young stars seem to be a common phenomenon. Supporters of extraterrestrial civilizations and seekers of contact with aliens are actively listening to the Universe. But as a first sensible step, we are searching for what we best know and understand: Earth-like life or remnants thereof, and hence we start by looking for water. Even in our local neighbourhood, in the Solar System, the space exploration that has been ongoing for

FIGURE 1.7 Water-ice on the Northern perennial polar cap of Mars. The image was taken with the camera of the Mars Global Surveyor spacecraft in 1999. For colour version, see plates section. (Image courtesy of NASA/JPL.)

more than half a century has brought us evidence for past liquid water on Mars; complex hydrocarbons and prebiotic molecules in the interstellar medium, comets and Titan among others; and possibly liquid water oceans below the surfaces of some outer satellites.

Indeed, as we will discuss thoroughly hereafter, the exploration of our Solar System to date has eventually brought us the revelation that liquid water may well exist not only on the surfaces but also underneath the crust of several Solar System objects like Mars (Figure 1.7), Jupiter's moons, Ganymede, Europa and Callisto, or Saturn's moons, Titan and Enceladus (see Chapter 4). All of these environments are only accessible through sophisticated investigation by close-in remote or *in situ* space missions, and therefore remain largely mysterious, but could well represent habitable worlds.

In the rest of this book, after defining the aspects of habitability that we shall make use of, and describing the notions that prevail today in the search for habitable conditions in the Universe, we will focus on tangible astronomical facts that allow us to look at the celestial bodies in our Solar System and beyond, from the perspective of hospitable locations for human life.

# 2　What is life and where can it exist?

The search for habitable worlds in the Universe entails our understanding of the conditions in which life appeared, survived and developed on Earth. This understanding has been growing consistently since the first geological, atmospheric, oceanographic and biological studies. As stated in *The Limits of Organic Life in Planetary Systems*, put together by the Committee on the Origins and Evolution of Life of the National Research Council (NRC, 2007):

> it is now clear that although terrestrial life is conveniently
> categorized into million of species, studies of the molecular
> structure of the biosphere show that all organisms that have been
> examined have a common ancestry. There is no reason to believe, or
> even to suspect, that life arose on Earth more than once, or that it
> had biomolecular structures that differed greatly from those shared
> by the terrestrial life that we know of today.

Our planet is not blessed everywhere with conditions favourable to human life, but in spite of the harsh and extreme chemical and temperature ranges that living species have to deal with, we have proof today that life thrives on Earth wherever liquid water and energy sources are available. However, other lifeforms may well exist, as has been suggested by some scientific studies. In what follows in this chapter we try to give an overview of terrestrial life and what it requires, touch upon other possibilities and focus on the environmental conditions necessary for the sustainability of life of the standard definition (Earth-like), before we begin our trip across the Solar System and elsewhere in quest of habitable places.

## 2.1    THE CONCEPT AND CONDITIONS OF LIFE

The Earth was born 4.56 billion years ago. Lost in some nondescript suburb of the Milky Way, this piece of rock was about to witness a unique phenomenon: life. Our planet hosts more than 8 million species today. Exactly how and when this proliferative life appeared remains a mystery in spite of regular discoveries and progress made on the subject in the past years, as we shall see hereafter. A few theories have been established based on observational evidence, numerous hypotheses and scarce facts. Chemists, biologists, geologists and astrophysicists pursue this essential human quest by exploring the depths of the oceans, the ancient continental rocks and interstellar space to seek traces of primitive life, habitable conditions and a better understanding of the chemical reactions that gave birth to living organisms.

Before describing these searches, we should perhaps begin by characterizing the terrestrial life that we know well, first through its macroscopic visibility and then through microscopic observation, which began in earnest about four centuries ago. Terrestrial life has basic ingredients that constitute its fundamental building blocks: carbon, nitrogen, hydrogen and oxygen.

### 2.1.1    *The building blocks of life*

Prior to the notion of organic chemistry, 'elements' had been defined as earth, wind, water and fire by the classical Greeks, or as gold, silver, mercury, etc. by the alchemists in the middle ages. Matter is composed of basic structural units that we call atoms (from the Greek ἄτομος, meaning that which cannot be divided) and that can in fact be divided into subatomic particles named protons, neutrons and electrons which interact with each other, bonding to form molecules. Some molecules are formed through sharing of the available electrons, and we call this 'covalent bonding'. When transfer of electrons from one atom to the other occurs, we have an 'ionic bond'. When the electrons are distributed equitably among the atoms, there is no charge – as for example in carbon dioxide, because carbon and oxygen have very similar

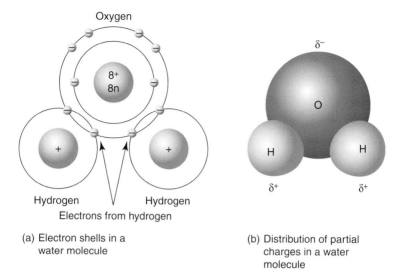

(a) Electron shells in a
water molecule

(b) Distribution of partial
charges in a water
molecule

FIGURE 2.1   Schematic drawing of the water molecule. (Adapted from Sam Adam-Day, http://alevelnotes.com/?id=135)

electronegativities and an even distribution of electrons. A polar molecule, on the other hand, is any molecule that has an uneven distribution of electrons and hence a charge. For example, water (which as we have seen already is essential for terrestrial life) is a polar molecule, because the oxygen atom is more electronegative (pulls electrons towards itself) than the two hydrogen atoms. The oxygen attracts the electron from each hydrogen atom, giving it a slightly negative charge, thereby leaving the hydrogen atoms with a slightly positive charge. Water's unique structure, with its oxygen and two hydrogen atoms (Figure 2.1), allows the two hydrogens to bond with the oxygen atoms of other water molecules, something that has significant consequences for its liquid and ice forms. It is responsible for the high temperatures required to make water boil or melt, because it is difficult to break the multiple hydrogen bonds and the special hexagonal lattice structure of the ice. Unlike other similar substances, water is very viscous and dense so that the solid phase (the ice) floats in the liquid.

The eighteenth-century French nobleman Antoine-Laurent de Lavoisier is rightly believed to be the 'father of modern chemistry'. Lavoisier studied chemistry and biology. At a time when water and air were still considered 'elements', he determined their composition, identified oxygen (Greek for producer of acid), hydrogen (Greek for producer of water) and nitrogen (*azote* in French, meaning lifeless in Greek). He also predicted the properties of silicon. With the help of his wife, Marie-Anne Pierrette Paulze, Lavoisier published in 1789 the first list of 23 elements in his *Traité élémentaire de chimie* (Elementary Treatise on Chemistry), which is actually the first textbook in modern chemistry. In it he introduced the principle of conservation of mass and provided the definition of an element as a substance that chemical analysis could not subdivide into lesser components, which led to the theory of the formation of chemical compounds from such elements. Lavoisier's experiments ultimately led to our perception of organic chemistry, including, for instance, the ordered periodic classification of 67 elements by Dmitri Mendeleev. Quantitative spectroscopy completed the laboratory experiments and allowed chemists and astronomers to work hand in hand in the early twentieth century to identify the spectral signatures of the gases of different elements and molecules observed in stars and planets. As we shall see hereafter, this remains the most reliable and fruitful means of investigation of the nature of planetary and exoplanetary objects.

We know today that the Universe is essentially composed of hydrogen (75 per cent) and helium (about 24 per cent) atoms, the simplest elements in the periodic table. The laboratory experiments on elements initiated in the eighteenth century led to the concept of organic chemistry, a branch of chemistry that concerns components which include the element carbon. Carbon-based molecules constitute the foundation of life on Earth. Thus, unlike most of the material in the Universe, typical living organisms are composed mainly of carbon, hydrogen, oxygen and nitrogen (CHON).

However different the composition of the Universe and of life may appear today, 13.7 billion years ago all existing atoms and elements had a common origin through the major event known as the 'Big Bang'. In the early stages of the expansion that ensued, when the Universe was just 3 minutes old and very small, collisions between protons produced hydrogen which then converted into helium (for about a quarter of the mass of the universe). Heavier elements were formed later through stellar nucleosynthesis, albeit in much smaller quantities. Living species are composed of only about 26 of the 92 naturally occurring elements, and 6 of those 26 make up practically all of their weight: oxygen (~70 per cent), carbon (~15 per cent), hydrogen (~10 per cent), nitrogen (~4 per cent), calcium (~2 per cent) and phosphorus (~1 per cent). The other 20 elements essential for life are present in very small amounts, some in such tiny amounts that they are designated simply as 'trace elements'.

All of these constituents combine to compose life as we know it and are studied in organic chemistry. The number of different molecules that are based on carbon is infinite. During the first half of the nineteenth century, chemists thought that only living organisms would produce organic compounds. This was later proven to be false, since organic compounds, hydrocarbons and their derivatives can be synthesized in a laboratory through abiotic (unrelated to life) processes. In 1828, for instance, F. Wöhler managed to produce carbamide (the organic chemical urea, a component of urine) without the 'vital force'. But organic chemistry remained separate (albeit with quite some overlap) from inorganic chemistry, which is a field dealing with the properties and reactivity of all chemical elements but which today is fundamentally directed more towards understanding the role of metals in biology and the environment, and studying materials useful in energy and information technology, or in other new sciences such as nanotechnology.

The atomic building blocks mentioned above and the interactions among them are responsible for all the biological molecules that go on to compose lifeforms. Biomolecular chemistry is a part of organic chemistry dealing with biological molecules which form the foundation of terrestrial life and come in a great variety of monomers and polymers. They include the following types of molecules:

lipids (hydrocarbons and acids) form membranes and store energy;

sugars (monosaccharides and polymerized sugars) ensure support and
energy storage;

amino acids, polymerized amino acids and proteins have many roles,
including supporting the metabolism and biological catalysis
accelerating metabolic chemical reactions;

polymerized sugars, phosphates and nitrogenous bases ensure the storage
and transfer of genetic information through two nucleid acids, DNA
(deoxyribonucleic acid) and RNA (ribonucleic acid). DNA contains
the genetic information while RNA can copy and transfer this
information.

An essential step in this chain of increasing complexity is the building
of boundary layers, or membranes, which allow for different types of
macromolecules to be differentiated and different functions to be
specified. How can membranes be formed? This is where liquid
water, as a dipolar solvent, plays a major role. Within a liquid water
solution, molecules have two types of reactions: the polar ones readily
dissolve and are called hydrophilic; non-polar ones have a low affinity
for water and are not easily soluble, and they are known as hydrophobic. Some molecular chains have both properties at once, with a
polar hydrophilic head and a non-polar hydrophobic tail. If such molecules (called amphiphiles) are immersed in liquid water, they tend to
align perpendicular to the water surface, with their heads in the water
and their tails in the air (Figure 2.2). This configuration creates a

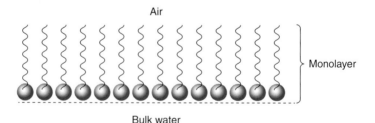

FIGURE 2.2  Diagram of a molecular monolayer. (Adapted from R. Torecki,
ILPI.)

monolayer of molecules – a membrane. In some cases, a double layer structure can be formed. The spontaneous formation of membranes in liquid water from a set of amphiphile molecules is actually the first step on the road to the development of cellular life.

## 2.1.2   Cells

It appears that life arose on Earth about 4–3.5 billion years ago. The earliest organisms identified in fossil remains were minuscule and rather featureless, and are almost impossible to distinguish from structures that originate through abiotic physical processes.

There is a consensus today among biologists that all living organisms on Earth must share a single last universal ancestor. It is considered virtually impossible for organisms to have descended from two or more different heritages that had evolved independently.

All living organisms are based on cells and depend on them for all their functions: absorbing nutrients, converting these nutrients into energy and reproducing. To help perform these functions, each cell stores its own set of instructions.

Until recently, there were two general categories of cells recognized: prokaryotes and eukaryotes. Prokaryotes are unicellular organisms that do not develop or separate into multicellular forms; eukaryotes include both unicellular and complex multicellular organisms. The simplest and earliest type of cells to evolve must have been prokaryotic organisms lacking a nuclear membrane, the membrane that surrounds the nucleus of a cell. Prokaryotes can inhabit most places on Earth, including on our own bodies and other surfaces. Bacteria are well-known prokaryotic organisms, continuously studied even today, some of which grow in filaments, or groups of identical albeit independent cells. Normally, there is no continuity or communication between these cells, although sometimes it may happen that they stay together (if for instance they did not separate after cell division). The recent discovery of a second group of prokaryotes, the Archaea, has added a new cellular branch of life and given us new insights into the origin of life itself. There are therefore three main

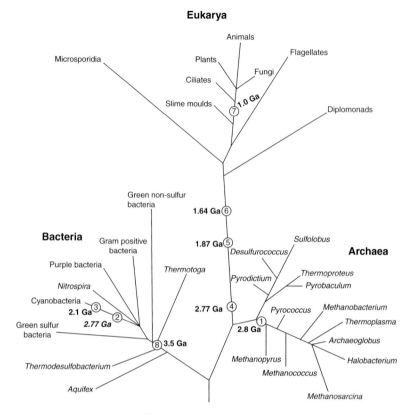

FIGURE 2.3  Tree of life. (Image from Brocks *et al.*, 2003.)

domains of life recognized today: Archaea, Bacteria and Eukarya (Figure 2.3).

The word Archaea comes from the Ancient Greek ἀρχαῖα, meaning 'ancient'. Among microbial lifeforms that are likely to have evolved at an early stage in Earth's history, and, consequently, are candidates for early emergence on other bodies in the Solar System and in exoplanetary systems, Archaea have for the past 30 years been identified as a separate taxonomic group. They display relatively simple growth requirements and are thought to be evolutionary relics. Although their early history is unclear, studies suggest that Archaea are genomic hybrids, containing features of information transfer

Table 2.1 *History of life origin and evolution on Earth*

| Life-related processes | Age (in billion of years ago, Ga) | Eon | Era |
|---|---|---|---|
| Internal structure of the Earth: in the centre an iron core surrounded by a silicate layer; the magnetic field is created by the movements of the magnetic core; the surface of the planet is a magma ocean; water, initially in vapour form, condenses in liquid oceans; the most ancient known rocks discovered in Canada date back to 4.2 billion years ago; cooling of the surface, oceans, atmosphere | 4.6–4.0 | **Hadean**: formation of the Sun in 100 000 years in the centre of a collapsing interstellar cloud; formation of the Solar System and its planets; a few tens of million years are enough for the formation of the Earth; Moon formed by impact | |
| Controversial discovery of biological activity found in rocks in Greenland, earliest evidence of life on Earth – prokaryotic bacteria dominate; most ancient stromatolites discovered | 4.0–2.5 | **Archaean**: Late heavy meteoritic bombardment | |
| Earliest multicellular organisms, first eukaryotic cells found; elevation of the oxygen amounts in the atmosphere due to the activities of living organisms, notably the cyanobacteria | 2.5–0.5 | **Proterozoic**: oxygenation of the atmosphere | |
| Nucleated cells arise; earliest known fungi | 2.5–1.6 | **Palaeoproterozoic**, continuation of the evolution of the atmosphere | |
| | 1.6–1.0 | **Mesoproterozoic** | |
| | 1.0–0.5 | **Neoproterozoic** | |
| Multicellular eukaryotes arise; earliest land invertebrates and plants appear followed by verterbrates and dinosaurs; explosion of biological diversity | 0.5–today | **Phanerozoic** | |

systems similar to higher organisms, and having metabolic capabilities similar to bacteria. Methanogenic Archaea (methanogens are micro-organisms that produce methane as a metabolic byproduct in anoxic conditions) can grow in the most minimal components (small molecules such as $H_2$ and $CO_2$ or CO), and are capable of growth in an extremely wide range of temperatures, from negative temperatures to as high as 122 °C. The methanogens are interesting microorganisms, as we shall see when we consider planetary objects where methane plays an important role.

Other organisms are multicellular, or have many cells – such as humans ourselves, composed of many cells (some estimates go up to a hundred thousand billion). Eukaryotes are thus multicellular, more evolved organisms which include protozoa, algae, fungi, animals and plants (Figure 2.4). A eukaryotic cell is roughly ten times as large and a thousand times greater in volume than a prokaryotic one. A eukaryotic cell contains a plasma membrane, glycocalyx components, a cytoskeleton and cytoplasm, in which we find small bodies called organelles. Most important among these is the distinctive nucleus, which is made of protein and deoxyribonucleic acid (that we commonly call DNA; Figure 2.5) and which is organized into units called chromosomes. Genes are small parts of these chromosomes and hence of the DNA.

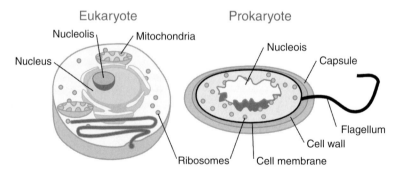

FIGURE 2.4 Cell types: (left) eukaryote, (right) prokaryote. For colour version, see plates section. (Image from the Science Primer, a work of the US National Center for Biotechnology Information, part of the National Institutes of Health.)

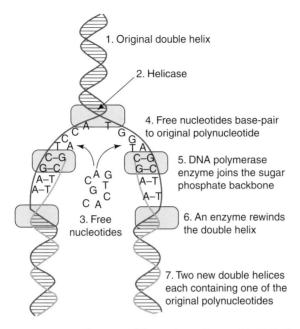

1. Original double helix

2. Helicase

4. Free nucleotides base-pair to original polynucleotide

5. DNA polymerase enzyme joins the sugar phosphate backbone

3. Free nucleotides

6. An enzyme rewinds the double helix

7. Two new double helices each containing one of the original polynucleotides

FIGURE 2.5   Schematic of deoxyribonucleic acid (DNA), showing the double standard helix with the genetic information encoded as a sequence of nucleotides (guanine, adenine, thymine and cytosine) recorded using the letters G, A, T and C. (Adapted from Burrell, 2002.)

The appearance of the eukaryotic cell was a milestone in the evolution of life. Prokaryotes and eukaryotes share the same genetic code and metabolic processes, but eukaryotes have a significantly higher level of organizational complexity that we think has led to the development of multicellular organisms populating the Earth today in vast numbers, and including plants, birds, fish and mammals such as humans.

It has been proposed that the oldest confirmed evidence of life on Earth, found in fossilized bacteria, dates back by about 3 billion years (gigayears ago, or Ga). Some other finds in rocks, thought to be bacteria, could be dated to about 3.5 Ga while some geochemical evidence (not yet confirmed) seems to indicate the presence of life as early as 3.8 Ga. However, non-biological processes might also be responsible for these supposed 'signatures of life', and a biological origin cannot be proven. At all events, the lack of fossil or geochemical evidence for earlier and simpler types of organisms than bacteria has led people to believe that life either arose spontaneously on Earth (see the next section) or that it was transported here from elsewhere in the Solar System (panspermia theory).

### 2.1.3   Origin of life on Earth

The processes thought to have contributed to the origin of life on Earth are as follows: the presence of a prebiotic soup containing a large variety of organic compounds is mandatory, although we still do not know the exact primordial recipe that led to life. The 'primordial soup' may have been fabricated from components existing already on Earth, or have included exogenous supplies brought in by comets, asteroids, micrometeorites or interplanetary dust particles. Originally it was thought that the primitive secondary atmosphere (not the same as the atmosphere acquired by accretion, but formed rather by outgassing from the interior or by cometary impacts, for instance) was reducing and contained mostly ammonia ($NH_3$) and methane ($CH_4$). However, the most current theories of the early Earth conditions suggest instead the presence of nitrogen combined with methane at the time of the Archaean (see Feulner (2012) for a review) as well as a hot and violent environment where the Earth was molten because of the immense heat caused by its accretion energy and by the volcanoes. Volcanoes ejected gases ($CO_2$, $N_2$, $H_2$) into the early atmosphere. In contrast with the case of the giant planets, hydrogen molecules are light enough to escape the terrestrial gravity field. Thus, it is likely that most of the atmospheric carbon was in the form of $CO_2$ with perhaps some CO, and the nitrogen mostly as $N_2$ (there was no $O_2$ present in the early terrestrial atmosphere; any outgassed $O_2$ released during the episode of melting and heating of the crust would have reacted with the metals of the crust, causing oxidation, while the lack of $O_2$ is crucial for the formation of organic molecules). It has been demonstrated in experiments trying to reproduce the Earth's original atmosphere (see Subsection 2.1.4) that in the absence of $O_2$, gas mixtures containing CO, $CO_2$, $N_2$, etc. produce the same results as those starting from $CH_4$ and

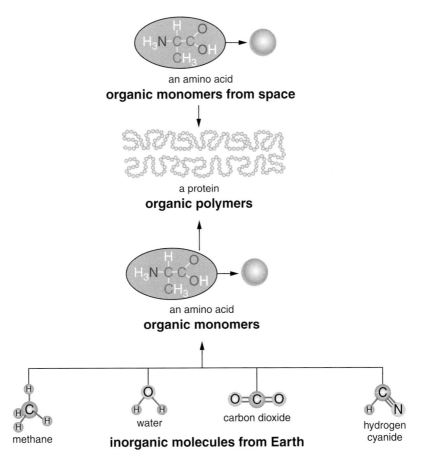

FIGURE 2.6 Prebiotic Earth with the sources for organic chemistry from exogenous and endogenous inputs. For colour version, see plates section. (Redrawn from an image by Jen Philpot and Jane Wang, courtesy of *Science Creative Quarterly*.)

$NH_3$. The hydrogen atoms are extracted essentially from water vapour, but gaseous mixtures with less hydrogen are more favourable to the development of aromatic amino acids under the conditions we expect of the primitive Earth.

At the same time, Earth was bombarded by asteroids, one of which, it is believed, dislodged the material that was to form the Moon. According to Jeffrey Bada and Gene McDonald (1996), a large variety of organic compounds, including those which play a major role in biochemistry such as amino acids, purines and pyrimidines, have been identified in one class of meteorites, the carbonaceous chondrites. Besides demonstrating that important biomolecules can be produced abiotically in extraterrestrial environments, their presence also suggests that exogenous compounds should be periodically delivered to the surface of the Earth (and other planetary bodies as well) by various processes.

Gradual cooling caused condensation of $H_2O$ to form rain and further reduced the temperature on the surface. Rainwater dissolved salts from the rocks, forming oceans. Since salts are held together by ionic bonds, chemical evolution may have begun in a salty ocean of pH 7.

Although theories and ideas about the atmospheric conditions and climate on the early Earth have made considerable progress in the past years (thanks to the work of, among others, James Kasting and co-workers), many uncertainties still remain. But the formation of an atmosphere containing $N_2$ and $CO_2$ gases and an ocean containing $H_2O$ appears to be a natural consequence of planetary accretion in the terrestrial planet region (Figure 2.7). Regardless of whether the important 'prebiotic soup' ingredients were produced locally or imported, the weakly reducing envelope forming around the Earth provided an environment favourable to the emergence and sustainability of life. As we have noted above, the current consensus is that the atmosphere contained little or no free oxygen initially, and that oxygen concentrations increased significantly about 2.0 billion years ago (Kasting, 1993), essentially as a consequence of photosynthesis and organic carbon burial. At the same time, the concentrations of carbon dioxide and other greenhouse gases declined so as to compensate for the increasing solar luminosity. Earth's relatively stable climate was probably a result of the negative feedback between atmospheric $CO_2$, surface temperature and the weathering rate of silicate rocks. Thus, it is also generally agreed that the atmospheric greenhouse effect must have been more important in the past, but the levels of carbon dioxide and other gases involved in the process vary according to the theories.

FIGURE 2.7 Liquid water seas on Earth; artist perspective by T. Encrenaz. For colour version, see plate section.

It is still uncertain how exactly the transition came about from the abiotic chemistry of the primitive Earth to the first self-replicating molecular systems capable of evolution. But the subsequent evolution of these first self-replicating molecules on our planet then gave rise to the RNA world and subsequently the DNA/protein world characterizing all life today.

Any investigations as to whether a planet hosts organic molecules should start by checking for compounds that are easily synthesized under plausible prebiotic conditions, or that are abundant in carbonaceous meteorites, and should focus on those that have an essential role to play in biochemistry. Amino acids satisfy all these criteria and should therefore become a focus for such investigations, even though we cannot be certain that amino acids were a component of the first self-replicating systems, or even an essential ingredient for the origin of life. Nevertheless, amino acids, synthesized in high yields in prebiotic experiments (see next section), are one of the more abundant types of organic compounds present in carbonaceous meteorites and are the building blocks of proteins and enzymes.

First to propose the idea of chemical evolution in 1923 were two geneticists, Alexandre Oparin (1894–1980) from Russia and John Burdon Sanderson Haldane (1892–1964) from the United Kingdom. The theory is based on two components.

> *Pattern component*: Increasingly complex carbon-containing molecules are formed in the atmosphere and ocean of the primitive Earth.
>
> *Process component*: Radiant and kinetic energy are converted into chemical energy in the bonds of large molecules.

The theory predicts a four-step process:

1. Small molecules containing carbon, such as formaldehyde ($H_2CO$) and hydrogen cyanide (HCN), are formed first.
2. The small molecules react with each other, to produce sugars, amino acids and nitrogenous bases, and hence form the prebiotic soup.
3. The small molecules in the prebiotic soup link together to form nucleic acids and proteins.

---

**BOX 2.1   Stromatolites**

In the earliest fossil records of life on Earth we find stromatolites, dating back to more than 3 billion years. These are colonial structures formed by photosynthesizing cyanobacteria (probably responsible for the creation of Earth's oxygen atmosphere) and other microbes. Stromatolites are, then, microbial communities, which have the capability to grow using only minimal chemical components, utilizing the energy of sunlight and the ability to fix $CO_2$ (the 'dark reaction' of photosynthesis). They used to thrive in warm aquatic environments, building reefs much as coral does today. They are present in environments including saline, hypersaline and anoxic (sea water or fresh water depleted of dissolved oxygen) areas. They are categorized as prokaryotes (primitive organisms lacking a cellular nucleus). The photosynthetic activities of such populations contributed significantly to changing Earth's atmosphere from reducing to oxidizing, altering the course of evolution and leading to the development of higher organisms.

4. One of these molecules becomes capable of self-replicating, becoming the first living organism, and marking the transition from the end of chemical evolution to the beginning of biological evolution.

Step 1 of the chemical evolution scheme involves the presence of organic chemistry on a planet. This is the case today on several other bodies besides our own planet, as we will examine in detail later in the book. Indeed, organics can be found in the atmospheres of the giant planets and more importantly their satellites, for example HCN on Titan and in the plumes of Enceladus. In this book we will concentrate more on this first part of the theory, as it is the more easily detectable in the cosmos, but hereafter in this chapter we describe briefly what is involved in steps 2–4.

### 2.1.4 Experiments on life: laboratory synthesis of amino acids

We now know in which environments organic molecules can be found; the next step is to understand how these molecules can be transformed into more complex molecules such as amino acids which can be used for the building blocks of life.

Step 2 of the chemical evolution theory is thus more complicated as it requires the formation of the prebiotic soup, containing amino acids, sugars and nitrogenous bases.

Amino acid molecules are composed of carbon, hydrogen, oxygen and nitrogen (CHON) and structured with an amine group, a carboxylic acid group and an additional side-chain that is specific to each amino acid. For example, in its complete formula as used in biochemistry, an alpha ($\alpha$-) amino acid is written as $H_2NCHRCOOH$, where R stands for an organic substituent. Here the amino group is attached to the carbon atom immediately adjacent to the carboxylate group (the $\alpha$-carbon). Depending on which carbon atom the amino group is attached to, or which side-chain (R-group) is attached to their $\alpha$-carbon, different amino acids (of varying size) are formed.

Amino acids are important because they are now recognized as the building blocks of proteins, which are essentially linear chains of amino acids, linked together in different sequences. This is the reason for the large variety of proteins that exist in living organisms. Twenty amino acids are naturally incorporated into polypeptides and are called proteinogenic or standard amino acids. These 20 are encoded by the universal genetic code. Nine standard amino acids are deemed essential for humans because they cannot be created from other compounds by the human body, and so must be taken in as nutrients.

The idea of such an evolution emerged in the 1920s when Oparin and Haldane raised the idea that organic molecules could, through a series of chemical reactions, evolve into prebiotic molecules and then into microorganisms under conditions that favoured such reactions, and that these conditions could be expected on the primitive Earth.

This theory received strong support in 1952, when two American scientists, Stanley Miller (1930–2007) and Harold Urey (1893–1981), managed to synthesize amino acids in the laboratory from a reducing medium including hydrogen, methane and ammonia in the presence of water, submitted to a series of electric discharges over several days. The purpose was to verify that, by assuming several conditions that would have prevailed on the primitive Earth and introducing them into the experiment, one would be able to recreate the species responsible for the origin of life.

The water cycle and lightning effects in the atmosphere were simulated through evaporation and condensation of water. The liquid water was heated to induce evaporation, sparks were fired between electrodes to simulate lightning through the atmosphere and water vapour, and then the atmosphere was cooled again so that the water could condense, in a continuous cycle (Figure 2.8).

Quite rapidly, the mixture was found to have changed colour. The results, after one week, showed the presence of a deep red solution containing HCN and $H_2CO$, which was interpreted by the authors as an indication that organic preliminary molecules could have been formed in a reduced atmosphere. By the end of two weeks of continuous operation, Miller and Urey observed that as much as 10–15 per cent of the carbon within the system was now transformed into organic

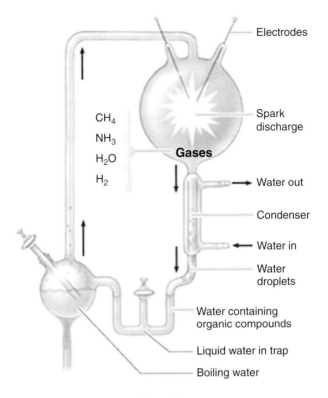

FIGURE 2.8 Schematic of the Miller–Urey experiment.

compounds. As noted by Barton *et al.* (2007), 'Two per cent of the carbon had formed the amino acids that are used to make proteins in living cells, with glycine as the most abundant. Sugars and lipids were also formed,' while nucleic acids were not found within the reaction. But most of the 20 common amino acids were there, in various concentrations. The experiments basically demonstrated that simple organic compounds and other macromolecules can be produced from a certain mixture of gases with the addition of energy.

Ultimately, all the amino acids used in the composition of living organisms were detected in this type of experiment. In later work, the radiation source was successfully replaced by an ultraviolet source, more compatible with the energy source likely to be present on the early Earth. The composition of the medium, however, was different from the putative primordial atmosphere of the Earth, which was probably dominated by carbon dioxide; laboratory experiments starting with an oxidizing medium led to poor results. Hydrogen, nevertheless, might have been present in the primitive atmosphere, as a result of a series of chemical reactions (the serpentinization process) between water and rocks below the surface, possibly also leading to the formation of methane and ammonia. In any case, the Miller–Urey experiment was a major milestone in our understanding of the role of prebiotic chemistry in the apparition of life.

After Miller's death in 2007, his former student, Jeffrey Bada, inherited the original equipment from the experiment. In the sealed vials, Bada and other scientists found that the experiment, already quite constructive and fruitful, had managed to produce even more different amino acids, at least 25 altogether and perhaps more. Miller and Urey's experiment should therefore be considered a remarkable success, because it managed to synthesize more complex organic molecules from simple chemicals than life itself really needs.

Additional experiments performed by Miller were subsequently re-examined – for instance one with conditions similar to those of volcanic eruptions, in which a nozzle sprayed a jet of steam at the spark discharge. In the reanalysis (Johnson et al., 2008), which used high-performance liquid chromatography and mass spectrometry, the scientists again managed to detect more organic molecules than Miller had originally. They also noted that the volcano-like experiment had produced, all in all, 22 amino acids, 5 amines and many hydroxylated molecules, more than any of the other experiments. The group suggested that this could explain why terrestrial volcanic island systems became rich in organic molecules, and also that the presence of carbonyl sulfide in these systems may have helped these molecules to form peptides.

The Miller–Urey experiment had been preceded by some other attempts using electric discharge (such as those by Wollman M. MacNevin, who never published results he thought to be 'too complex for analysis'; and by K. A. Wilde, who used voltages up to only 600 V on a binary mixture of carbon dioxide ($CO_2$) and water in a flow system, leading to only small amounts of carbon dioxide being transformed to carbon monoxide, and finding no other significant reduction products or newly formed carbon compounds). It was to inspire many more. In 1961, Joan Oró found that the nucleotide base adenine could be made from hydrogen cyanide (HCN) and ammonia in a water solution. His experiment produced a large amount of adenine, the molecules of which were formed from five molecules of HCN (Oro, 1961). Many amino acids can also be formed from HCN and ammonia under these conditions. Experiments conducted later showed that the other RNA and DNA nucleobases could be obtained through simulated prebiotic chemistry with a reducing atmosphere (as found in later steps of the chemical evolution). HCN is a molecule that is quite common in the Universe, and in particular it exists on Saturn's moon Titan.

Although there are theories that support the $NH_3 + CH_4$ assumption, we do not really know whether the early Earth atmosphere was actually reduced, but we think this was the case since the high volcanic activity introduced gases such as carbon dioxide, nitrogen, hydrogen sulfide ($H_2S$) and sulfur dioxide ($SO_2$) into the Earth's early atmosphere around 4 billion years ago through volcanic eruptions. When these gases were added to the ones in the original Miller–Urey experiment, a larger range of molecules was found. In the presence of oxidized gases, water does not yield HCN, $H_2CO$, or other small carbon-containing molecules. Bada noted recently that in current models of early Earth conditions, reactions between carbon dioxide and nitrogen produce nitrites, which destroy amino acids as fast as they form (Bada, 2004). On the other hand, the early Earth may have contained enough iron and carbonate minerals to neutralize the effects of the nitrites. Bada further reported that with the addition of iron and carbonate minerals, the experiment yielded large quantities of amino acids, something that would also have occurred on Earth even with an atmosphere containing carbon dioxide and nitrogen.

In conclusion, the Miller–Urey experiment shows that formaldehyde and HCN can form under a reduced environment, but – as we mentioned above – the Earth's early atmosphere may well have been oxidized, not reduced. In that case the origin of the chemicals in the early Earth environment remains a mystery. However, we must note that most of the natural amino acids, hydroxyacids, purines, pyrimidines and sugars necessary for life, or variants thereof, have been found in the Miller–Urey experiment.

Akiva Bar-Nun and Sherwood Chang studied an oxidized atmosphere in which they injected sunlight-like radiation by means of a lamp. They used water vapour and an oxidized gas, carbon monoxide (CO). Their experimental results showed that a wide variety of reduced-carbon compounds ($H_2CO$, acetaldehyde and $CH_4$) had formed. When the temperature conditions and the types and proportions of the gases were changed, similar results were obtained. This experiment supported the hypothesis that sunlight can trigger the reduction of carbon from a mixture of volcanic gases (in Step 1 of chemical evolution).

All of the aforementioned evidence and observations (some of which can be found in the recommended reading lists) support the occurrence of amino acids in the prebiotic soup. Since amino acids formed readily in Miller and Urey's spark-discharge experiment, they could have also formed in the conditions of early Earth. Alternatively, amino acids may have been delivered from outer space, for instance in the interstellar dust that sprinkles the Earth and contains hydrogen cyanide and aldehydes, key reactants in forming amino acids. Therefore, amino acids could have formed in the ancient oceans or been dropped into it by meteorite impacts.

Simulations conducted in 2005 at the Universities of Waterloo and Colorado suggested that the early atmosphere of Earth could have been composed of up to 40 per cent hydrogen, which would have constituted a more hospitable environment for the formation of prebiotic organic molecules. If the escape rate of hydrogen was low, its presence in the planet's early atmosphere would have induced the production of organic compounds more easily, again lending weight to the scenario of the primordial soup. The Waterloo/Colorado group conducted some outgassing calculations using a chondritic model for the early Earth composition and found that they supported the scenario of a reducing atmosphere on the early Earth, re-establishing the importance of the Miller–Urey experiment (Fitzpatrick, 2005).

In the Solar System, one can look for bodies, such as the Murchison meteorite (found in Australia near Murchison, Victoria, in 1969), where one could imagine the Miller–Urey experiment taking place. Analyses did indeed show that more than 90 different amino acids were contained within the Murchison meteorite, 19 of which are components of terrestrial life. Besides meteorites, comets and other icy outer Solar System bodies are thought to contain large amounts of complex carbon compounds (such as tholins) formed by these processes, darkening the surfaces of these bodies. The early Earth also is thought to have been heavily bombarded by comets, possibly thus acquiring a large supply of complex organic molecules along with the water and other volatiles that they injected. This has been used to suggest an origin of life outside of Earth: the panspermia hypothesis (see Box 2.2).

While most scientists consider the formation of sugars and the origin of amino acids to be reasonably well understood, unresolved problems in laboratory simulations of early Earth remain. They include questions that underline our ignorance of many aspects, such as how and why RNA was able to form, given that different pentoses and hexoses form in approximately equal amounts, whereas the formation of RNA needs ribose to have been dominant.

Among current investigations in laboratories, attempts are being made to probe the limits and tolerances of human life, as well as to define the relevant physiological and genomic mechanisms operating under stress and under the extreme conditions found elsewhere in the Solar System.

## 2.1.5 Chirality and the specificities of human life

The term chirality is derived from the Greek word (χειρ) for hand. It is a scientific basis for the layman's concept of 'handedness'. Indeed, human

BOX 2.2   **Panspermia**

The panspermia theory suggests that life on Earth did not originate on our planet, but was transported here from somewhere else in the Universe, most likely the Solar System. While this idea may seem far-fetched, some considerations suggest that an extraterrestrial origin of life should not be immediately dismissed.

The earliest known advocates of the panspermia theory were the ancient Greek philosophers Anaxagoras and Socrates, but the idea became superseded by Aristotle's preferred theory of spontaneous generation. Closer to the modern era, in the nineteenth century, the French chemist Louis Pasteur, the British mathematician and physicist Lord Kelvin and others argued that life could originate from space. This was supported by the Swedish chemist Svante Arrhenius who declared that bacterial spores, propelled through space by the pressure of light, had originally brought life to Earth.

One more recent scientific argument that supports the panspermia theory is the emergence of life soon after the heavy bombardment period of Earth, between 4 and 3.8 billion years ago. During this period, researchers believe the Earth endured an extended and very powerful series of meteor showers. However, the earliest evidence for life on Earth suggests that it was present some 3.83 billion years ago, overlapping with this bombardment phase. These observations suggest that living things during this period would have been threatened by extinction, supporting the idea that life might not have originated on Earth (or would have disappeared even if it had emerged) but might have been brought in from outside at a more favourable time.

However, in order for life to have originated elsewhere in the Universe or even the Solar System, there would have to be an environment on another planet capable of supporting it. In other words, one would need a habitable planet close by. Our study of the Universe suggests that terrestrial life would have a hard time surviving outside the Earth.

In the 1920s, separate studies published by the Russian biochemist Alexander Oparin and British geneticist J. B. S. Haldane produced a mixture of all theories in which the doctrine of spontaneous generation

BOX 2.2  **(cont.)**

was revived in a more sophisticated form. In the new version, the spontaneous generation of life no longer happens on Earth, cannot be reproduced in a laboratory because it takes too long, and may well have left no signs of its occurrence behind. As described in the main text, in 1953, American chemists Stanley Miller and Harold Urey showed that some amino acids can indeed be chemically produced from ammonia and methane in the laboratory.

Then in the 1970s, British astronomers Fred Hoyle and Chandra Wickramasinghe revived interest in panspermia by observing, in the dust between the Earth and distant stars, evidence for traces of life. This discovery was soon questioned, however, as other scientists pointed out that these signatures were also found in abiotic organic matter. Hoyle and Wickramasinghe also proposed that comets, which are largely made of water-ice, could carry bacterial life across galaxies in their interiors, thus protecting it from radiation damage and extreme conditions during its travel.

Another theory, put forward by British chemist James Lovelock in the 1970s, suggested that from its early stages life influences the environment in which it tries to evolve, to make it more and more suitable for life's welfare. This theory, which William Golding named Gaia, has gained some followers, although it refutes the idea of Darwinian evolution. Facing strong criticism, Lovelock has retreated slightly from some of his earlier bold claims for Gaia.

All in all, while scientists accept today that some of the ingredients for life may exist in space, there remains a lot of scepticism about the subject.

hands are a good example of chirality: the left hand is a non-superposable mirror image of the right hand. No matter what the orientation of the hands, one cannot make the two hands coincide (Figure 2.9).

The term chiral in general is used to describe an object that is not superposable on its mirror image (Nic *et al.*, 2006). Achiral ('not chiral') objects are objects that are identical to their mirror image.

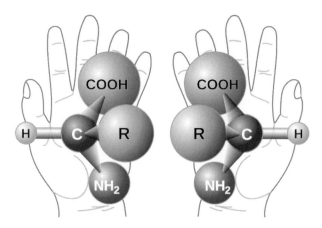

FIGURE 2.9  Two enantiomers of a generic amino acid superimposed on human hands. (Image courtesy of Wikimedia Commons, http://en. wikipedia.org/wiki/File:Chirality_with_hands.svg.)

In chemistry, chirality usually refers to molecules. In Paula Bruice's book *Organic Chemistry* (2009) a chiral molecule is defined as a type of molecule that has a non-superposable mirror image, and the feature that is most often the cause of chirality in molecules is found to be the presence of an asymmetric carbon atom. Chiral molecules that are mirror images of one another and are non-superimposable on one another are called enantiomers or optical isomers. Pairs of enantiomers are often designated as 'right-' and 'left-handed' (represented as L for *laevo*, or left; D for *dextro*, or right).

Several of the molecules considered as 'active' from the biological point of view are chiral, including the amino acids (building blocks of proteins) and the sugars in DNA. In biological systems, most of these compounds are of the same chirality: the amino acids are mostly L (and usually have no taste), and the sugars are usually D (and taste – as expected – sweet). As a consequence, left-handed proteins are made of L-amino acids, whereas D-amino acids produce right-handed proteins.

At one time, chirality was associated with organic chemistry, but this misconception was contested by Alfred Werner, a Swiss chemist, who resolved in the laboratory hexol, a purely inorganic compound.

But the chirality concept is still an excellent tool for distinguishing between a biological and a non-biological origin of some species. Enzymes made up of all D-amino acids function just as well as those made up of only L-amino acids, but the two enzymes react with the opposite stereoisomeric substrates (reactants that are identical in atomic constitution and bonding, but different in the three-dimensional spatial arrangement of their atoms). There are no bio-chemical reasons why L-amino acids would be favoured over D-amino acids.

Therefore, a major unresolved question remains: in the amino acid chemical evolution, why did only left-handed enantiomers survive? Although two forms of every amino acid (except glycine which has no enantiomers) exist in nature, only the left-handed ones are found in living beings. A related question of equal importance refers to the emergence of solely right-handed sugars during the chemical evolution.

The origin of this homochirality in biology is the subject of many studies and much debate. It is likely that terrestrial life's 'choice' of handedness was purely random, and that if carbon-based lifeforms exist elsewhere in the universe, their chemistry might, in theory, have opposite chirality. There is some suggestion, however, from some data analyses, that if amino acids formed in comet dust at early stages, circularly polarized radiation (which makes up 17 per cent of stellar or solar radiation) might have led to the selective destruction of one of the two forms of chirality of amino acids, producing a bias which ultimately resulted in all life on Earth having the same chirality.

To test this hypothesis, laboratory experiments have been conducted to study the interaction of circularly polarized ultraviolet (UV) radiation with icy molecules analogous to those found in comets or the interstellar medium. In particular, a team from Orsay University, in France, using light produced by the SOLEIL synchrotron, has been able to show that more enantiomers of one type than the other were produced by circularly polarized radiation rather than by using linearly polarized radiation. This difference in production was measured in alanine, $CH_3CH(NH_2)COOH$, the simplest chiral proteic amino acid, and is compatible with the excess found in meteorites. The result may have an important astrophysical implication. High rates of circularly polarized radiation have been found in the much more extended region of star formation in the Orion Nebula. It is also important for life on Earth, because if amino acids were racemic (meaning they had equal proportions of both enantiomers), they would probably not have been efficient catalysts in early living organisms.

Enzymes and their substrates (reactants) can be chiral because all except one of the amino acids from which they are composed are chiral. If one considers the analogy of hands and gloves, the glove could be the enzyme into which the hand (a substrate) is inserted: it fits well only in one case (the right-hand or the left-hand glove). If the substrate does not fit properly, the enzyme will not be able to act on the substrate. This is why an enzyme will act on one enantiomer more readily than on another.

Understanding the events that led to the origin of life on Earth is complicated by the lack of geological evidence around 4 billion years ago (4 Ga) when the transition from prebiotic chemistry to biochemistry is thought to have taken place. Although erosion and plate tectonics have erased the terrestrial geological records from this time, there is a possibility that information about this period of Earth history may be still preserved on Mars, which lacks plate tectonic activity, and has suffered far less surface alteration. Indeed, if amino acids are found everywhere on the surface of the Earth, they might also exist on the surface of Mars, where extensive areas of the surface may date to more than 4 Ga, independently of their origin (abiotic or prebiotic). In addition, if similar to terrestrial life, proteins and enzymes related to life on Mars might have also been subject to the amino acid homochirality principle, although it is possible that Martian life could be equally based on D- or L-amino acids.

Geomorphological evidence suggests that liquid water existed on the Martian surface at some point in the past, because early Mars may have had a more humid, denser atmosphere than today, possibly comparable to that of the early Earth (although probably less dense; see Subsection 3.3.3). If this is the case, then Mars could have gone through at least some of the steps leading to the origin of biochemistry like that on Earth. This is why current exploration of Mars focuses on the search for traces of prebiotic chemistry, or any biochemical evidence that could be interpreted as extinct Martian biota. Although some advocate that life may even still exist on Mars today, in some protected subsurface environments, and that the exploration of the red planet should include such investigations, most scientists agree that it is unlikely in current conditions.

And indeed, even if they existed, searching for amino acids in foreign environments like Mars is no easy task. Because of contamination from Earth components, only an upper limit of 0.1 part per billion (ppb) could be inferred for the amino acid AIB found in samples of the lunar soils (AIB is 2-aminoisobutyric acid, also known as α-aminoisobutyric acid, α-methylalanine or 2-methylalanine, an amino acid contained in some antibiotics of fungal origin, rather rare in nature). If we really want to detect traces of amino acids in samples returned from Mars or other places, we will need large enough samples to avoid contamination from terrestrial amino acids or any other life-related constituents.

### 2.1.6 Another diagnostic for recognizing living matter: isotopic ratios of carbon

Another measurement can be used to identify biogenic material in a source outside the Earth: it is based on the measurement of carbon isotopes.

As discussed above, the abundances of elements and isotopes in the Universe are, to first order, those measured in the Sun, and they are expected to broadly reflect the so-called 'cosmic abundances', because they are the product of primordial and stellar nucleosynthesis. There are departures from this rule, however, in particular for isotopic abundances. Isotopic fractionation, as it is called, can originate from different processes: temperature change, condensation/evaporation or biological processes.

On Earth, the $^{12}C/^{13}C$ ratio measured in minerals is equal to 90. However, in living organisms, this ratio is altered as a result of photosynthesis. Plants preferentially use the lighter isotope ($^{12}C$) when they convert sunlight and carbon dioxide into organics. There are two photosynthesis cycles being used on Earth: the C3-photodissociation cycle, used in temperate environments, is the most common and leads to a change in $^{12}C/^{13}C$ by 2.6 per cent, while a smaller change is observed in the C4-photodissociation cycle, favoured in hot and dry environments. As a result, the $^{12}C/^{13}C$ ratio in living organisms is about 92.4.

Measuring the $^{12}C/^{13}C$ ratio in a carbonaceous material on an extraterrestrial body would thus, in principle, provide us with a test of its abiotic or biogenic nature. The measurement is a priori feasible from remote-sensing spectroscopy, but reaching the per cent level precision would be difficult. In the case of Mars, the Mars Science Laboratory (Curiosity) rover is equipped with a tunable laser spectrograph capable of measuring $^{12}CH_4$ and $^{13}CH_4$ separately, provided the methane abundance is above a few tens of ppb. Unfortunately, the first analyses in the search for methane on Mars have been unsuccessful so far (see Subsection 3.3.3).

## 2.2   DEFINITION OF LIFE AND HOW TO LOOK FOR IT OUTSIDE ITS USUAL ENVIRONMENT

As defined by NRC (2007), 'terrestrial life then uses water as a solvent; It is built from cells and exploits a metabolism that focuses on the carbonyl group (C=O); it is thermodynamically dissipative, exploiting chemical-energy gradients; and it exploits a two-biopolymer architecture that uses nucleic acids to perform most genetic functions and proteins to perform most catalytic functions.'

And space exploration to date, conducted by humans, naturally tends to search for Earth-like lifeforms, thus focusing on places where water and energy are available. 'Carbon chauvinism' is a term coined by the famous planetary scientist Carl Sagan, according to which, 'it is not surprising that carbon-based organisms breathing oxygen and composed of 60 per cent water would conclude that life must be based on carbon and water and metabolize free oxygen' (Sagan, 1973). Some people, however, do manage to think otherwise, and scientists have advanced biochemical theories according to which life could form based on ammonia, hydrocarbons or silicon, and solvents different than water could exist (see Subsection 2.2.3). Stephen Hawking, however, once noted in a conference lecture on 'Life in the Universe' that in spite of these possibilities, 'carbon seems the most favorable case, because it has the richest chemistry'.

The natural tendency toward terracentricity (believing that the Earth is in the centre of the Universe) requires that we make a conscious effort to broaden our ideas of where life is possible and what forms it might take. The long history of terrestrial chemistry tempts us to become fixated on carbon because terrestrial life is based on carbon. But basic principles of chemistry warn us against this trend. It might be possible to conceive of chemical reactions that might support life involving non-carbon compounds, occurring in solvents other than water, or involving oxidation–reduction reactions without dioxygen.

In the quest for life inside and outside our Solar System, the question asked is not only how we define life but also whether we could recognize it if we happened to run into it. Although this book is not about the search for living organisms (the reader is referred to books in Further Reading on that subject), we try to give hereafter some examples of specific criteria that would have to be met in order for us to definitively identify living organisms. More essentially, we try to characterize which places could have been life-friendly today or at some point of their evolution and might be suitable for sustaining life, once it has appeared or been brought into the environment.

### 2.2.1 Can we completely define life?

In spite of intense data gathering, speculation, modelling and experimentation for generations, the definition of life has eluded scientists and philosophers. It may in fact not be feasible to try to define such a concept. The Committee on the Origins and Evolution of Life of the National Research Council (NRC, 2007) lists some of the characteristics of life as we know it on Earth:

- It is chemical in essence; terran living systems contain molecular species that undergo chemical transformations (metabolism) under the direction of molecules (enzyme catalysts) whose structures are inherited, and heritable information is itself carried by molecules.
- To have directed chemical transformations, terran living systems exploit a thermodynamic disequilibrium.
- The biomolecules that terran life uses to support metabolism, build structures, manage energy, and transfer information take advantage of the covalent bonding properties of carbon, hydrogen, nitrogen, oxygen, phosphorus, and sulfur and the ability of heteroatoms, primarily oxygen and nitrogen, to modulate the reactivity of hydrocarbons.
- Terran biomolecules interact with water to be soluble (or not) or to react (or not) in a way that confers fitness on a host organism. The biomolecules found in terran life appear to have molecular structures that create properties specifically suited to the demands imposed by water.
- Living systems that have emerged on Earth have done so by a process of random variation in the structure of inherited biomolecules, on which was superimposed natural selection to achieve fitness. These are the central elements of the Darwinian paradigm.

Some other properties of terrestrial life are given by Mike Muller of the University of Illinois (Muller, 2011) and in the textbooks recommended therein. They include order, reproduction, growth and development, energy use, homeostasis and evolutionary adaptation.

> *Growth and development.* In order to ensure that the species survives and flourishes, the organisms transmit their characteristics
>
> *Energy utilization.* All terrestrial living organisms require energy to function, and they find it in different forms that they transform and adapt to what is required for them to subsist
>
> *Order.* If enough energy is available (in the form of solar radiation, geothermal processes etc.), the living species use the resources and organic components available in their environments and redistribute

them according to their needs. In this way, the system as a whole functions properly, ensuring its evolution.

*Reproduction.* Procreation (sexual or asexual) is a fundamental characteristic of life without which a species becomes extinct. Some organisms have higher yields of reproduction than others, and the efficacy of the process depends on the food availability, the climatic conditions and other parameters.

*Homeostasis.* From the Greek ομοιο meaning 'the same' and στασις meaning 'stable', homeostasis is a property of living organisms that ensures that their internal systems remain stable within normal functioning limits in spite of any changes of the external environment. This capacity to adapt is essential for the survival of the species.

*Evolutionary adaptation.* As the organisms adapt to their changing environment, they evolve; and their physical or biochemical characteristics, inscribed in their genes as well as affected by the environment, are modified to enhance biological complexity.

## 2.2.2 Extreme conditions on Earth today

Terrestrial life, as we defined it before, with carbon compounds and water as solvent, has to abide by strict constraints imposed by our chemistry, and must manage to emerge and survive within such boundaries as 'the boiling and freezing points of water, acid and alkaline extremes, the presence or absence of oxygen, and other factors', to quote Penelope Boston (2010).

Our understanding of these constraints and the way that terrestrial life has to push these boundaries has expanded greatly in only the past 30–40 years. The search for living organisms that survive and even thrive in extreme environments on Earth has brought us new insights as to the possibilities of life elsewhere in the cosmos and the form it might take.

When we try to define an 'extreme' environment, we can imagine places where one would be exposed to drought, excessive heat or cold, large radiation doses or lack of nutrients, and where any living organisms would have to adapt in order to survive. For instance, there are ways for some species to bypass the stringent temperature conditions in which liquid water can exist. As Boston points out, some organisms and microorganisms, like the tiny crustaceans known as 'water bears', use highly concentrated sugars, salts and amino acids in their cells to survive in sub-freezing temperatures. We call such species 'extremophiles'. Earth offers all sorts of extreme niches for these organisms, ranging from the superheated waters of submarine volcanic vents to the freezing Antarctic Dry Valleys. One has to realize that more than 80 per cent of our planet – far from being a balmy paradise – is at temperatures colder than 5 °C; but even these places are inhabited. Indeed, some organisms such as the icefishes of Antarctic waters cannot stand warm temperatures and are limited to living in cold waters. In contrast, other species manage to make their homes in environments where high pressures depress the physical boiling of the liquid and can thus survive at high temperatures. A microbe called 'Strain 121', for example, survives at 121 °C, and we believe this is close to the highest temperature that life can withstand on Earth.

However, there are limits to what proteins and DNA can sustain. If a protein loses its three-dimensional structure and becomes 'de-natured', it loses its function. Factors that can cause this denaturing include changes in acidity (pH), the salt concentration, the presence of reducing agents, and extreme temperatures (in particular higher temperatures can reduce the strength of hydrogen bonds). Indeed, the chemistry and adaptations of creatures like the Antarctic icefishes requires them to live at cold temperatures. Above 4 °C, they suffer from 'heat exhaustion' and may die.

An example of the possible synthesis and survival of complex organic molecules was found in the so-called 'black smokers' or hydro-thermal vents in the deep oceans (Figure 2.10), characterized by high pressures and superheated water (up to 450 °C). The reaction of cold sea water on dissolved compounds of iron, sulfur, nickel and reduced carbon induces some sort of chemical evolution leading the hydro-thermal vents to emit methyl mercaptan ($CH_3SH$) which combines with carbon monoxide to form acetic acid ($CH_3COOH$), a building block of organic molecules.

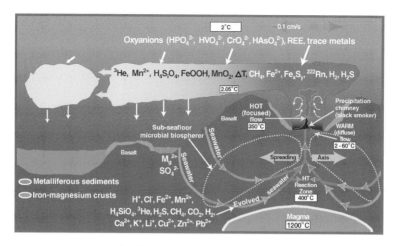

FIGURE 2.10 Hydrothermal vents and black smokers diagram with the biogeochemical cycle. For colour version, see plates section. (Image courtesy of Wikimedia Commons, http://en.wikipedia.org/wiki/File: Deep_sea_vent_chemistry_diagram.jpg.)

Some of the earliest forms of life found on Earth, the Archaea that we discussed in Subsection 2.1.2, were originally perceived as extremophiles that lived only in difficult conditions in superheated waters, frozen Antarctic ice and hypersalted lakes. To survive at very low and very high temperatures (above 100 °C), Archaea have developed enzymes that hold their cells together. However, we now know that Archaea actually also exist in much more normal conditions, in ordinary temperatures in our body, and in the ocean plankton. They are now recognized as a major and essential component of Earth's life and may even have an active role in the carbon and the nitrogen cycle.

In organisms more complex than microbes, their behaviour can overcome the physical constraints. For complex organisms it is of course possible to use retreat as a defence mechanism, and so life just relocates in some cases. In the desert, some animals develop diurnal habits by burying themselves into the more humid under-surface layers to avoid the burning sun. Researchers returning from Antarctica reported the presence of some nematode worms (unsegmented worms, with a long rounded body with pointed ends) and other organisms that can stand the harsh cold temperatures by producing antifreeze and the lack of water by drying themselves out and letting the wind drift them around until water comes back. This strategy of sitting tight and waiting for the conditions to improve is a good one for the survival of many organisms.

Are there environments on other worlds like these extreme conditions or even harsher, but in which we could find life? And how far can we push the connection between Earth's extreme environments and life on other planets? We use our understanding of our own planet's most challenging environments to look for clues to the most life-hospitable conditions on other planets, and hints of what we should be searching for when we investigate those other worlds for signs of life.

Certainly there are Earth environments that resemble some features of the other worlds in the Solar System in their lack of water, energy, warm temperatures and nutrients. For instance, we know that some microbes survive in the Antarctic Dry Valleys, even in frozen lakes. Could the waterless surface of Mars hide such living organisms, or could we envisage a habitat deep beneath the surface on the red planet? At the very least, we may find fossils of past living beings on Mars. In Earth's deep oceans, around hydrothermal vents, life does not simply exist but in some cases thrives without the solar radiation that we view as so indispensible for life. Similar conditions prevail in the subsurface oceans of some of the giant planets' satellites (Europa, Ganymede, Titan. . .); given enough time, then life might also develop there. We do not know precisely how long it takes for life to emerge and evolve, but some proposed theories ('punctuated equilibrium', for instance) argue for a rapid evolutionary pace.

We know that some of the extreme environments on Earth have persisted for millions of years. There is even evidence suggesting that life on this planet originated in very inhospitable conditions – possibly a hot, sulfurous and non-oxygenated environment that we would now consider very extreme!

For example, the hypersaline Antarctic lakes or the marine sediments are hosts to organisms resistant to high levels of radiation and

FIGURE 2.11 Río Tinto, a river with extreme acidic living conditions where some Earth organisms still survive, thought to have similarities with some places of subterranean Mars. For colour version, see plates section. (Image credit: J. Segura and R. Amils.)

low temperatures (below 0 °C). Some are anaerobic (from the Greek word αναερόβιος, which literally means 'living without air', as opposed to aerobic); others grow by feeding on hydrogen and carbon dioxide.

Examples can be found in the Río Tinto in southwestern Spain that flows down from the Sierra Morena mountains of Andalusia to the Gulf of Cádiz at Huelva (Figure 2.11). This river has attracted significant interest among the scientific community and in particular among astrobiologists because of the discovery of extremophile aerobic bacteria that live in its water and are considered to be responsible for its high acid content. The bacteria rather surprisingly feed on the subsurface rocks on the iron and sulfide minerals found in the river bed. The extreme conditions in this river have been hypothesized to be analogous to other locations in the Solar System thought to contain liquid water, such as subterranean Mars or Europa (some scientists have made direct comparisons between the chemistry of Río Tinto's waters and these environments).

Habitats of this kind need more investigation. But we also need to expand our methodology for looking for life, because we may still be missing quite a few environments on Earth that harbour life while being representative of conditions on alien worlds. And the places where no life has ever emerged on Earth also require investigation in order to better determine the constraints and the limits.

Within the Solar System, the best candidates for supporting life beyond Earth are Mars, Titan, Europa, Ganymede and Enceladus; the latter four are icy moons around giant planets, a long way from the Sun. The temperatures in their neighbourhood are very low (50–100 K) and therefore the conditions are different from the ones on Earth. Furthermore, except for Titan, none of these bodies possess substantive atmospheres as a protection against solar and cosmic radiation; indeed, Europa is well within the Jovian magnetosphere. Consequently, insensitivity to radiation is likely to be a requirement for survival in those environments.

As said at the beginning of this Section, learning about extreme conditions on Earth can help us plan ahead for our search of habitats in the Universe. With that prospect, when we explore the extreme conditions under which life exists on Earth we need to ask ourselves the questions:

1. If other life forms are possible that are still based on carbon and need liquid water, can they differ in some of their other properties such as reproduction, homeostasis or polymeric structures?
2. Might there be living organisms that do not require water? For example, can polymers or even monomers support catalysis and genetics in non-aqueous environments, using solvents other than water that are available on other Solar System bodies?
3. Can the basic structures and functions of life – for instance, those that allow interactions between minerals and organics, and those that lead to evolution – still operate in non-aqueous media, especially those found on Solar System bodies other than the Earth?

Antibiotic resistance, as found in bacteria, is thought to be largely due to so-called 'horizontal' or 'lateral' gene transfer, which is an ancient and efficient process for rapidly creating diversity and complexity. Another property of Earth's organisms is 'sociability' which allows them to live in communal fashion among their own species, to use their genes in original ways and to change the genes they have if that allows them to better adapt to environmental conditions.

Now that we have identified on Earth a wide range of extremophiles or organisms that can survive in conditions that may approach those found in other planetary bodies (significantly higher or lower pressures, temperatures, atmospheric density etc.), we could still greatly benefit from:

- considering available information on how life adapts to extreme environments, including low temperature, high radiation, high salt and high desiccation, and derive the implications for the prospects for life on Mars, Europa and other icy moons and exoplanets. This will also directly bear on potential forward contamination of these bodies by microbes that could be present on spacecraft from Earth (see Chapter 5).
- examining the potential of circular polarization as an unambiguous biosignature, sensitive to life at the broadest level, from simple microbial species to complex terrestrial scenes.
- using Solar System objects as reference cases for characterization tools that can be incorporated in planetary probes, and have the potential to be adapted for use in the next generation of space telescopes designed to study the detailed characteristics of terrestrial and gas giant exoplanets.

### 2.2.3 Other possible forms of life

Because we have only one example of biomolecular structures that adapt and surmount life's constraints, and because the human mind stumbles at imagining lifeforms completely different from those it recognizes, it is difficult for us to assess how life might present in environments different from our own planet.

Some of the requirements for life appear to be stricter than others. For instance, carbon is a necessary element for biomolecules as we know them. And biology tells us that water is the most efficient solvent for life, because it is required by DNA. Nonetheless, scientists have imagined molecular structures adequate for life but very different from our terrestrial ones, and probably difficult to detect.

To quote planetary scientist Jonathan Lunine (2005) of Cornell University, 'Does life require liquid water as the liquid medium, or are other liquids possible hosts for, if not life as we know it, some kind of organized chemistry? You'd be testing the limits of what the word "life" really means in the cosmos.'

Macromolecules that use silicon do exist. Silanes, which are structural analogues of hydrocarbons, could be used as a building block for life under the right conditions. It is difficult, however, to understand how they could have emerged spontaneously to create a biosphere.

As the next step one can consider a set of observations about life that might be considered exotic when compared with human-like life. Life found on Earth has inhabited all places where thermodynamic disequilibria and water are available, and these living organisms are not so exotic that we cannot recognize their biochemistry or their ancestry.

The quest is on for environments in the Solar System that might be suitable for life of the terrestrial type. Our current understanding of the Solar System indicates that most sites are at thermodynamic disequilibrium, an absolute requirement for chemical life, and that many locales at thermodynamic disequilibrium also have solvents in liquid form (not necessarily water: Titan, for instance, has hydrocarbon liquid extents). Furthermore, environments exist where the covalent bonds between carbon and other lighter elements are stable. When all of these requirements for life appear together, it is tempting to consider the place suitable for life.

Scientists still wonder if some kind of biochemistry exists that could be adapted to those exotic environments, much as human-like biochemistry is adapted to terrestrial environments. Because few detailed hypotheses are available today, and we do not know if and at what stage the abiotic processes that manipulate organic material in a planetary environment could lead to life, most of what we have to work with is speculation.

Beyond our terracentric approach, it is also clear that using thermal and chemical energy to maintain thermodynamic disequilibria, covalent bonding among carbon atoms, water as the solvent, and DNA as a molecular system to support Darwinian evolution is not a unique way to create life. Synthetic biology is a new field that aims to invent and construct alternative biological entities such as enzymes, genetic circuits (DNA modules that perform logic operations in cells) and other biological systems. The field has already provided laboratory examples of new chemical structures that go through genetics, catalysis and Darwinian processes. Who knows what such experimental methods will provide in the future?

## 2.3   WHAT IS A HABITABLE ZONE (A HABITAT)?

In what interests us hereafter, we look at the environmental conditions recognized today by scientists working in various fields (biology, geology, physics, chemistry, etc.) as being mandatory for a living organism to be able to survive and develop. When a celestial body supports all of these conditions we call it a 'habitat'.

### 2.3.1   *Classical concept of the habitability zone*

Planetary habitability is the ability of a planetary environment to support and sustain lifeforms. The habitability potential of a planet or a satellite depends on a combination of factors, which are considered to be essential for the appearance, evolution and maintenance of life. Crucial factors are, among others, the orbital properties of the planetary body, its stability over long periods of time, its bulk composition, the existence of an atmosphere and a surface, as well as the proper chemical ingredients.

Life emergence on Earth sets the habitability constraints in our Solar System. As discussed earlier, terrestrial life is thought to be the final product of a long complex chemical evolution, requiring at least four raw ingredients: (a) the existence of liquid water, (b) a stable environment, (c) carbonaceous matter as nutrients and (d) energy. The fulfilment of these prerequisites over a long period of time can be considered as an indicator of suitable environments for hosting the

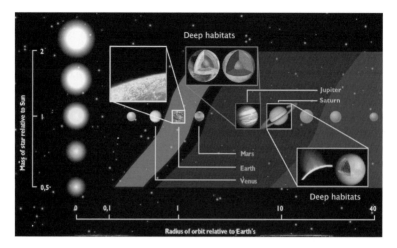

FIGURE 2.12 The habitable zone in our Solar System and elsewhere. For colour version, see plates section. (Adapted from Lammer *et al.*, 2009.)

proper biological building blocks, which may lead to the formation of primitive life structures.

When looking for habitable regions in the Solar System, we therefore consider primarily the areas around a star or a planet favourable for life in any form. These are called habitable zones (HZ). In Figure 2.12, the habitable zone (light blue) is plotted as a function of the spectral type for main sequence stars. The planets of our Solar System are indicated. The 'centre' of the HZ is defined as the distance a planet would have to be from its star in order to receive enough energy to be at the right temperature, pressure and luminosity conditions to allow water to remain in a stable liquid state on its surface. In such a zone, living organisms may arise and evolve or at least survive if transported there. The term 'Goldilocks planet' (from the tale in which a little girl has to pick just the right items among a bear family's possessions) is attributed to all planets located within the circumstellar habitable zone (CHZ, also known as the ecosphere; thought to exist around all stars, not only the Sun) which present exactly the 'right conditions' for life to appear, being at the right place at the right time. Goldilocks planets are obviously of key interest to scientists looking either for existing (and possibly intelligent) life or for future homes for the human race, as we shall see in Chapter 6.

For instance, liquid water is recognized as the best solvent for life to emerge and evolve, as we saw in the previous section. Water is quite abundant in the cosmos, but it only exists in liquid form at temperatures from 273 K (0 °C) up to values which depend on the ambient pressure; on Earth the maximum temperature for liquid water to exist is 373 K (100 °C). We have seen already that even on Earth, the temperature surface conditions vary greatly. In general, we can consider a range between –230 °C and +300 °C in the Solar System depending on whether we are looking at far-away Pluto or at Mercury, critically close to the Sun. On exoplanets even larger ranges of temperatures can exist.

The term 'snow line' or 'ice line' refers mainly, in stellar systems, to the distance from the star beyond which some compounds – water, but also ammonia and methane – can exist only in ice form. No liquid water is expected on the surfaces of bodies located beyond the snow line, which in the case of the Solar System is at around 5 AU for average temperatures around 150 K.

Habitable zones are not set regions in the stellar systems. As the star lives its life and goes through different phases, the nature of the zone moves and changes. The stars considered as right for habitable zones to develop around them, like our Sun, a G2-type star, are the middle-class stars in terms of mass and luminosity: 'early F' or 'G' to 'mid-K' types. The associated spectral ranges are in the 4000–7000 K range. In contrast to the upper-class luminous main-sequence stars like 'O', 'B' and 'A', which have short lives (about a few billion years), these average and unexceptional stars have the advantage of being long-lived, so that life has a better chance to emerge and evolve. Furthermore, even more usual 'K' type, fainter stars and 'M' red dwarfs are long-lived and could allow for habitable zones with planets to form. To date, only about a dozen planets have been confirmed in the habitable zones of stellar systems other than our own, but the exploration continues apace, and the Kepler mission has reported that very large numbers of such planets exist in our Galaxy.

The distances from these stars at which liquid water can exist are not close enough for a planet to be necessarily in tidal locking, but still not too far away, so that the ultraviolet radiation can still play a major role in favouring important atmospheric processes like ozone formation without being destructive for life. Planets 'tidally locked' to their star means that they rotate around themselves once per their year, or a fractional number of times per year. The Moon is tidally locked to the Earth and rotates once per month. Tidal locking, like for Mercury and the Sun, means that one hemisphere of the planet would permanently face the star and the other hemisphere would always be in the cold. Furthermore, objects in the habitable zones of smaller stars would be likely to experience larger tides that could remove axial tilt, resulting in a lack of seasons which would gradually remove the water from the planet.

Another problem for habitable planets located in the HZ around 'M' and 'K'-type stars is that they are usually close to their parent star and are therefore exposed to adverse effects such as damaging tidal forces and solar flares. Even if life originated on such a planet, it would evolve differently from Earth, as the atmosphere and in some cases magnetospheres would experience extreme long-term stellar radiation and plasma exposures. Thus, a combination of stellar and geophysical conditions can prohibit the existence of life on planets within the HZ. Space weather, in particular stellar radiation and stellar variation, can significantly affect the ability of planets within the HZ to retain surface water. Venus and Mars are examples of planets that may have experienced large-scale and relatively rapid loss of surface water, whereas our own planet possesses a set of defence mechanisms against the effects of space weather. Indeed, a combination of magnetosphere, atmosphere, geological and geophysical cycles may be essential to sustain stable bodies of water on the surface of planets, as it does on the Earth.

On our planet, the protective atmosphere is generated and replenished by volcanism and other processes such as the carbon cycle or biological effects. Different processes have been observed on other planets, such as the exchange of materials between Enceladus, the rings and Saturn (through the jets that bring out water and organics through Enceladus' south pole geysers), as well as between Io, Jupiter and its other moons. On Triton, Titan and Enceladus, as on some other bodies, outgassing of methane and/or some type of cryovolcanism in general are some of the other theoretical processes that have been evoked. Cryovolcanism happens when water, laden with other volatiles and organics, is ejected through a process similar to volcanism on Earth by geysers or other edifices, essentially owing to internal heating.

So even if living organisms appeared during the early stages of a planet's formation, the space weather effects could in the end lead to inhabitable and even hostile lifeless environments like those of Venus or Mars. This may appear to be in favour of the theory of interstellar 'dead zones' where life cannot exist, supporting the Rare Earth Hypothesis.

Most of the time and up to recently, when used in the context of planetary habitability, the Rare Earth Hypothesis implies terrestrial planets with conditions roughly comparable to those of Earth (Earth analogues). The boundaries of our own habitable zone in the Solar System, according to the traditional definition mentioned above and requiring the existence of liquid water on the surface, extend from 0.95 AU to 1.2 AU. However, these boundaries are flexible. The outer boundary of our HZ has been estimated to be at more than 2.4 AU, where we could have an environment with $CO_2$ ice clouds in the atmosphere allowing for the surface temperature to be above the freezing point of water. In current literature, one can find arguments as to why our HZ should extend somewhere between 0.675 and 3 AU.

Furthermore, habitable conditions can be found not only on the surfaces of Earth-like planets: a subsurface ocean within the satellite of a gas giant for example may be habitable for some lifeform similar to or very different from the Earth paradigm. Indeed, icy surfaces may cover liquid oceans, and move and fracture through plate tectonics redistributing the internal material and energy through an interconnected system. Our search for life in exotic habitats of the Solar System, such as Mars, Europa, Titan and Enceladus, and the discovery of planets beyond, means that habitability needs a better and broader definition.

### 2.3.2   *Extension of the habitable zone*

At a given moment, planets are in equilibrium with their surroundings: they are neither getting hotter nor colder. All planets absorb the incident radiation from the Sun that does not get reflected, helping them keep warm, and radiate the same amount of energy out into space in order to maintain equilibrium. The temperature of a planet can be approximated by assuming that it is a black body (meaning a surface that absorbs all radiant energy falling on it).

However, as stars evolve on the 'main sequence' (a special period in the evolution of a star), their luminosity and temperature increase, pushing the habitable zone outwards with time. At the beginning of the Solar System's formation, some 4.5 billion years ago, some studies suggest that our Sun was 30 per cent dimmer than it is today in its middle age.

Also, planets are not ideal black bodies. Water vapour and other gases are opaque in the near-infrared (where the maximum of the black body emission would be) so that in fact planets are hotter than the equivalent black body. This again pushes the habitable zone outwards.

On the other hand, the habitable zone is also more extended inwards because not all the incident sunlight is absorbed: some of it is reflected. The albedo of a celestial body is the fraction of sunlight reflected with the respect to the amount received and is quite high for Earth and Venus thanks to the presence of clouds.

These considerations influence our perception of habitability for the terrestrial planets, our immediate neighbours, as follows. Space weather can negatively affect the capability of planets to maintain liquid water inside a habitable zone subject to the influence of magnetospheres, stellar radiation or luminosity variations. In particular, Earth's siblings Venus and Mars, have experienced several different effects such as trapping of radiation by greenhouse gases, interaction with the solar wind and photodissociation which led to the loss of their hydrosphere.

Without protection by such properties as an atmosphere or a magnetic field, the habitability of a planet is in danger. An atmosphere can help to regulate the thermal conditions on a planet and hence help it keep its water in liquid condition. However, sometimes the chemical content of an atmosphere, such as the large amounts of $CO_2$ on Venus, contribute to a runaway warming (called a runaway greenhouse effect) so that all water that could possibly have been present on Venus at the beginning is now lost after it was quickly vaporized and dissociated by solar UV radiation in the upper atmosphere (see Section 3.2). When the temperatures consequently rose, the hydrogen escaped to space, breaking the water cycle. As a consequence, Venus is located today just at the inner threshold of the habitable zone, at 0.72 AU, while at the early stages of the Solar System it could well have been situated in a more favourable environment for habitability.

Similarly, runaway cooling can occur on a planet, such as a process possibly responsible for episodes known as Snowball Earth, where much of the planet's surface is hypothesized to have frozen over.

Mars orbits today at 1.52 AU, and scientists tend to believe that in the early times of the Solar System, between 3.5 and 4 billion years ago, the red planet was warm enough to allow for the presence of liquid water on the surface. But that water was lost, for reasons explained in detail in Section 3.3 but essentially due to extreme solar wind and ultraviolet conditions caused by the absence of a magnetosphere.

But in addition to the standard considerations, one should take into account the fact that liquid water need not be restricted to the surface. Liquid water can also exist in subsurface environments on planets or satellites inside and outside the habitable zone, like on Mars, but most significantly on Jupiter's Europa, Ganymede and Callisto, and Saturn's Titan and Enceladus. Therefore, the determination of the habitable zone limits can be expanded. The question is whether these underground water deposits can support any lifeforms.

In addition, although many organisms live in water, life would probably never have spontaneously originated and evolved in a body of pure water. This is because, as we said in *Life on Earth and Other Planetary Bodies* (Coustenis *et al.*, 2012):

> while there are many organisms living *in* water, none we know of is capable of living *on* water alone because life requires other essential elements such as nitrogen and phosphorus in addition to hydrogen and oxygen. Moreover, no known organism is made entirely of water. 'Just water' is therefore not an auspicious place for the emergence and evolution of life.

These subsurface oceans within the gas giants' icy satellites may be in direct contact with heat sources below their icy crust, or may be encapsulated between two ice layers, or as liquids above ice. If this situation persists for long timescales, the liquid underground ocean may become capable of sustaining life. Similar conditions have been considered for the primitive Earth where a coupled sea/ice system could provide the necessary conditions for life to emerge.

Evidently a new classification for the habitable types of celestial bodies is necessary. One such was proposed in a recent review article by Helmut Lammer and co-authors (Lammer *et al.* 2009; Figure 2.13). In this classification, 'Class I [are] habitable planets where complex multi-cellular surface life forms as we know on Earth can evolve [and which] also need to orbit around the right star' (such as G-type stars, and K and F-types with masses close to those of G stars). Stars of this type develop habitable zones at great enough distances that their stellar activity, which also tends to decrease rapidly, does not prove dangerous for any habitable planets emerging there. As long as the planet is active with plate tectonics of periods of a billion of years or more, then it can also preserve its atmosphere and liquid water and become habitable.

Class II habitats are environments where life may originate and evolve, but the atmosphere and magnetosphere environments of these planets, located within the HZ of low-mass M- and K-type stars and

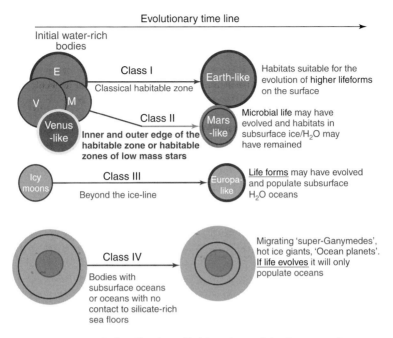

FIGURE 2.13 A classification of habitats in our Solar System and elsewhere. (Adapted from Lammer *et al.*, 2009).

therefore very close to their host stars, suffer extreme stellar radiation and plasma exposures over very long periods of time, possibly most of their lifetime. These are the cases we discussed previously, where in spite of their being in the HZ, thermal and non-thermal atmospheric escape processes modify the atmospheres and liquid water content of the planets so that they become inactive after some hundreds of million years and end up as dry Venus-like or cold Martian-like planets. For the Class II habitats, there is uncertainty as to whether life, having emerged in the early stages of the planet's formation, might have benefited from favourable conditions long enough to allow for an evolution that might persist even after the loss of (almost) all water, as in the case of some of the extremophiles we discussed in Subsection 2.2.2. In that case, the idea of panspermia leads to the possibility that this life might even have been able to propagate through space to another planet, independently of whether it survived in its original abode. But for any kind of life to have evolved beyond the simplest forms, the right environmental conditions have to be maintained for long periods of time.

In Class III habitats, subsurface oceans are in contact with silicates on the sea floor. In such environments on Earth, reduced radicals such as $H_2$ are contained in the hot fluid and can provide energy for a variety of organisms. Without similar reduced radicals, the source of energy necessary to power an organism on other planets could be a problem.

Class IV habitats can be found in the Solar System but also in exoplanets where an internal water ocean is possible between two thick layers of ice or different liquids, so that the delivery of external material may become feasible in the case of porous ice, or even better, a liquid layer.

While in the cases of Class I and Class II habitable planets, which will be discussed in Chapter 3, the main question is whether favourable conditions persist long enough to allow for life to develop and evolve after it gets started, the question for the Class III and Class IV habitats is more whether life could start at all, as we shall see in Chapter 4.

Nevertheless, icy moons in the Solar System and exoplanets may host the right ingredients for developing (today or in the past) life-friendly conditions, so the interest in an astrobiological investigation of these planetary objects is enormous. Since 2004, the discoveries of the Cassini–Huygens mission in the Saturnian system have revolutionized our considerations as to whether these bodies could harbour life of their own or at least provide us with valuable information on the origin and evolution of life on Earth.

Some of the processes mentioned above that are harmful or even fatal to life, in addition to the absence of tectonic or volcanic activity, tidal effects and other sources of energy such as radioactive decay, may affect the conditions for supporting life on a planetary body. And while it is possible that terrestrial life could adapt to an environment like that of Titan or Mars (see the 'terraforming' concepts in Chapter 6), it is far less likely that life could emerge there in the first place. It follows that a planet that has moved out of a habitable zone is more likely to host life than one that has moved into it.

There are good reasons today to believe, with all the new discoveries from space missions looking at exoplanets, that habitable zones can exist around many other stars, some of which may not offer the same conditions as those in our Solar System, but rather make available other solvent constituents (such as ammonia or methane). Atmospheric pressure conditions and greenhouse effects can also enable planets with special conditions to sustain surface water. And just as different levels of atmospheric pressure can affect whether water remains in a liquid state, the presence of dissolved compounds such as ammonia or salts in water can lower its freezing point, thus extending the habitable zones. For example, water containing brine is proposed as an explanation for seasonal flows on some of the warmer Martian slopes. Astrobiologists are studying the possibility of alternative biochemistry in such environments.

### 2.3.3   Prebiotic chemistry

In our quest for extraterrestrial habitats, we have to consider places where carbon is present and could be involved in the formation of complex molecules. The interstellar medium is thus the first place to consider. Indeed, observations performed since the 1970s, especially with ground-based radiotelescopes, have demonstrated that the interstellar medium is rich in complex organic molecules. Over 120 gas molecules have been detected so far, including a large majority of carbon-bearing molecules (see Table 2.2); the most complex gas molecule detected in the interstellar medium so far is $HC_{11}N$. As pointed out before, the ability of carbon to form complex molecules is due to its four valences or bonding possibilities. No atom other than carbon is able to form such complex interstellar molecules; in particular the silicon atom, the next element with four valences in the Mendeleev table, has not been found in gaseous molecules with more than one Si atom. The reason is probably that the cosmic abundance of Si is more than ten times lower than that of carbon. This fact, independent of any

Table 2.2 *List of currently detected interstellar molecules*

| H species | Carbon chains and cycles | H, C and O species | H, C and N species | H, C, N and O species | S, Si and other elements | Deuterated species |
|---|---|---|---|---|---|---|
| $H_2$ | CH | OH | NH | NO | SH | HD |
| $H_3^+$ | $CH^+$ | $HCO^+$ | $NH_3$ | HNO | SiN | $DCO^+$ |
| | $CH_3$ | $HOCO^+$ | $CH_3NC$ | $N_2O$ | HF | HDCO |
| | $C_2H_2$ | $H_2COH^+$ | $CH_3C_3N$ | HNCO | OCS | $DC_3N$ |
| | $l\text{-}C_3H$ | $CH_3CHO$ | CN | $NH_2CHO$ | $H_2CS$ | $H_2D^+$ |
| | $c\text{-}C_3H$ | $CH_2CHOH_3$ | $HCNH^+$ | $NH_2CH_2COOH$ ? | C5S | $N_2D^+$ |
| | $CH_4$ | $(CH_3)_2CO$ | $HC_3N$ | | CS | $D_2CO$ |
| | $C_4$ ? | CO | $HC_3NH^+$ | | SiO | $DC_5N$ |
| | $c\text{-}C_3H_2$ | $HOC^+$ | $C_2H_5CN$ | | AlF | $D_2H^+$ |
| | $l\text{-}C_3H_2$ | $H_2CO$ | $NH_2$ | | $HCS^+$ | HDS |
| | $C_4H$ | $CH_3OH$ | $H_2CN$ | | HNCS | HDCS |
| | $C_5$ | $c\text{-}C_2H_4O$ | $HC_2NC$ | | FeO | $C_4D$ |
| | $C_2H_4$ | $(CH_3)_2O$ | $C_5N$ | | SO | HDO |
| | $C_5H$ | $HOCH_2CH_2OH$ | $HC_7N$ | | SiS | $D_2S$ |
| | $l\text{-}H_2C_4$ | $CO^+$ | HCN | | CP | $CH_2DOH$ |
| | $HC_4H$ | $C_2O$ | HCCN | | $c\text{-}SiC_2$ | $CH_2DCCH$ |
| | $CH_3CCH$ | $C_3O$ | $NH_2CN$ | | $C_3S$ | CCD |
| | $C_6H$ | $CH_2CHO$ | $CH_3NH_2$ | | AlNC | $NH_2D$ |

| | | | | |
|---|---|---|---|---|
| C$_6$H$_2$ | CH$_3$OCHO | CH$_3$C$_5$N ? | SO$^+$ | CD$_2$HOH |
| HC$_6$H | CH$_2$CHCHO | HNC | HCl | CH$_2$DCN |
| C$_7$H | C$_2$H$_5$OCH$_3$ | C$_3$N | PN | DCN |
| CH$_3$C$_4$H | H$_2$O | C$_3$NH | SiCN | ND$_2$H |
| C$_8$H | CO$_2$ | C$_2$H$_3$CN | c-SiC$_3$ | CD$_3$OH |
| C$_6$H$_6$ | HCOOH | HC$_9$N | NS | c-C$_3$HD |
| C$_2$ | HC$_2$CHO | N$_2$H$^+$ | NaCl | DNC |
| CH$_2$ | CH$_2$OHCHO | CH$_2$CN | H$_2$S | ND$_3$ |
| CCH | CH$_3$CH$_2$CHO | CH$_3$CN | NaCN | CH$_3$OD |
| C$_3$ | HCO | HC$_5$N | SiH$_4$ | |
| | H$_3$O$^+$ | HC$_{11}$N | SiH | |
| | CH$_2$CO | CH$_2$NH | AlCl | |
| | C$_5$O | | C$_2$S | |
| | CH$_3$COOH | | MgCN | |
| | CH$_3$CH$_2$OH | | SiC$_4$ | |
| | | | SiC | |
| | | | KCl | |
| | | | SO$_2$ | |
| | | | MgNC | |
| | | | CH$_3$SH | |

FIGURE 2.14 The Horsehead Nebula. The dark region corresponds to dense molecular clouds where interstellar chemistry is active. For colour version, see plates section. (Image courtesy of Adam Block, Mt Lemmon SkyCenter, University of Arizona.)

anthropomorphic consideration, is a strong argument for the universal role of carbon chemistry in extraterrestrial life, if it exists.

Dense molecular clouds where the temperature is very low ($T$ = 10–20 K) are especially rich in complex organic molecules (Figure 2.14). In this medium, most of the matter is in the solid form. Interstellar grains are believed to be composed of a silicate core surrounded by a mantle of ices. Most simple molecules (composed of a few atoms only) are in solid form at these temperatures; two notable exceptions are $H_2$ and CO. Any other gaseous molecule encountering such a grain is immediately captured in the ice mantle. Under the effect of ultraviolet stellar radiation, organic molecules tend to evolve into more complex molecules. If subject to a denser and warmer medium (for instance the environment of a star in the process of formation), the complex molecules return to their gaseous form, thus enriching the ambient medium.

Another favourable environment is the envelope of carbon-rich evolved stars, where polycyclic aromatic hydrocarbons (PAHs) have been found in abundance. These molecules consist of benzene rings on which several tens of atoms (mostly hydrogen) are bound; in contrast with the smaller molecules, which are detected at radio wavelengths, they have been identified through their infrared emission, especially around 3 μm. When the envelope of such a star is ejected, these molecules feed the interstellar environment. Formamide, $NH_2CHO$, the simplest amide, was detected in 2013 in the gas surrounding the young Sun-like star IRAS 16293-2422, in the nebula of ρ Oph with the IRAM 30m radio-telescope (Spain). Formamide was already detected in the ISM, in Orion and Sagitarius B2, as well as in the coma of the Hale-Bopp comet. It seems to be as abundant as water. This could indicate a relationship between the chemistry at work in the environment of IRAS 16293-2422 and the chemical processes leading to the formation of our own solar system.

In spite of the success of the Miller–Urey experiment, we still do not know whether terrestrial amino acids were formed *in situ*. An alternative explanation assumes an external source. Meteoritic bombardment provides one possible solution to explain how extraterrestrial molecules might have been transported to Earth. The story of planetary formation tells us that an intense meteoritic bombardment (LHB, Late Heavy Bombardment) took place some 3.8 billion years ago, as observed from the crater-counting record at the surface of Solar System bodies. As was discussed above, this event was probably associated with the migration of the giant planets and the crossover of the Jupiter–Saturn system at the (2:1) resonance in their early history (see Section 1.2). The Spanish

FIGURE 2.15 The Murchison meteorite, in which a large number of amino acids have been identified. (Image: © New England Meteoritical Services, courtesy of D. Darling.)

biochemist Joan Oro (1923–2004) first suggested that the LHB event might have been responsible for bringing prebiotic molecules on Earth. This hypothesis received decisive support in 1969 with the fall in Australia of a primitive meteorite, known as Murchison (Figure 2.15). About 20 amino acids, including those found in proteins, were detected in this meteorite. There was a remarkable similarity between the amino acids found in Murchison and those found in the Miller–Urey experiment, as well as their relative abundances. This Murchison discovery was also a milestone in our attempt to decipher how life arrived on Earth. It demonstrated that prebiotic chemistry was already at work very early in the Solar System history, some 800 million years after the planets' formation. It also suggests that prebiotic molecules might have come on Earth from meteoritic impacts.

What is the origin of meteorites? Their parent bodies are believed to be of two kinds. The larger meteorites that arrive on Earth in solid form come from different types of asteroids; their mineral composition reflects the nature of their parent body. The most interesting of these, from an astrobiological point of view, are the primitive ones, called carbonaceous chondrites. The Murchison meteorite belongs to this category. Micrometeorites are able to penetrate the Earth's atmosphere without being totally destroyed and can be found in some privileged sites such as Antarctica (Figure 2.16). They are believed to be of cometary origin, which makes this material most precious. Indeed, comets are among the most primitive objects in the Solar System.

FIGURE 2.16 A micrometeorite collected in Antarctica. Comets are believed to be the parent bodies of micrometeorites. (Image credit: CSNSM/Orsay/ CNRS/IPEV, courtesy of Wikipedia.)

Comets are small bodies, only a few kilometres in size, and are mostly made of water and rocks. Their orbits are very eccentric, so that they spend most of their lives at large heliocentric distances, in a cold and rarefied environment, free from collisions and too small to be subject to thermal differentiation. As a result, comets can be considered as witnesses of the early processes which took place at the beginning of the Solar System history.

The primitive nature of comets was clearly demonstrated in 1986, at the time of the apparition of comet Halley. With a period of 76 years, this famous object, known since antiquity, was the object that allowed the English astronomer Edmund Halley (1656–1742) to identify the true nature of comets. Its 1910 apparition led to a ground-based observing campaign, which yielded beautiful, spectacular images. The 1986 apparition was less favourable in terms of geometric configuration (the comet being behind the Sun at perihelion), but the comet was approached by an armada of five spacecraft. In addition, infrared and radio measurements, not available at the beginning of the twentieth century, were used to determine the nature of the parent molecules that compose the nucleus. One of the main discoveries of the Halley campaign was the evidence for a close connection between interstellar and cometary matter, both in the gaseous and in the solid form. This conclusion was confirmed in 1997, with the apparition of another new bright comet, Hale–Bopp. Comets can thus be considered as possible ground-truth laboratories for prebiotic chemistry, much easier to study than the interstellar medium. A major question is: are there amino acids in comets? We hope to get the answer to this question when a European spacecraft, Rosetta, approaches a cometary nucleus, Churyumov–Gerasimenko, to probe its surface, interior and environment (see Section 4.4).

## 2.4  SEARCHING FOR EXTRATERRESTRIAL LIFE: FROM HABITATS TO CIVILIZATIONS

We focus hereafter on the conditions necessary for evolved forms of life as we know it to emerge or to be supported in extraterrestrial environments. We know today, based on our exploration, that any lifeforms we might find in the Solar System would be restrained, primitive and well hidden. For an astrobiologist (and for that matter, a planetologist, a geologist or a biologist, etc.) such a discovery would be a ground-breaking step in our understanding of life on Earth and therefore a major discovery. But of course, the ultimate goal for some people, when looking out in the Universe, is to seek other civilizations, similar to ours or more evolved, and popular fiction portrays aliens as powerful and intelligent (often hostile) entities. Scientists have tried to answer the question of the possibility of such extraterrestial worlds.

### 2.4.1   Could there be extraterrestrial civilizations?

First presented in 1961, the Drake equation assembles different factors used to estimate the number of extraterrestrial civilizations in our Galaxy, whose electromagnetic emissions are detectable. It is used, for instance, in the field of the search for extraterrestrial intelligence (see Chapter 6). The equation was devised by Frank Drake, then working as a radioastronomer at the National Radio Astronomy Observatory in Green Bank, West Virginia.

The Drake equation expresses the number of detectable civilizations in our Galaxy ($N$) as:

$$N = R^* \times f_p \times n_e \times f_l \times f_i \times f_c \times L$$

where the different factors express the following:

$R^*$:   the yearly averaged rate of star formation;

$f_p$:   the fraction of these stars that host one or more planets;

$n_e$:   among these planets, the average number that are habitable;

$f_l$:   the fraction thereof on which life does appear and develop;

$f_i$:   the fraction of the above where life evolves into intelligent beings and civilizations;

$f_c$:   the proportion of these advanced lifeforms that develop technology capable of transmitting a signal;

$L$:   the time for which such civilizations send detectable signals across space.

The Drake equation has no unique solution, and it is related to the Fermi paradox in that it tends to suggest that although a large number of extraterrestrial civilizations might form, the lack of evidence for such civilizations might mean that there is an additional factor that we have ignored (called the Great Filter) which actually leads to a much reduced value. This could be the case, for instance, if civilizations do not arise easily, or if any civilization that develops the technology needed to contact us would tend to be quickly extinguished (a foreboding idea for our own civilization).

Another far-fetched idea is that a superintelligent extraterrestrial life might exist: a much higher form of civilization than ours, but one which deliberately does not contact us so as to not interfere with the evolution and development of human life (the 'Zoo hypothesis').

It is clear that this 'equation' as written contains so much uncertainty in its terms that one cannot use it, as proposed on some websites today, to infer scientifically sound conclusions. Even if in our current era we are close to imagining that we could soon have some idea about the $f_p$ fraction, we are still far from knowing how often and how many civilizations appear and develop, and how long they last in the detected stellar systems, nor whether all of these civilizations originated around their current primary star or elsewhere, if interspace colonization is possible. However, with these thoughts, Drake did nevertheless lay the foundations for constructive discussions on the subject, in particular since the discovery of exoplanets.

Searches for signals from other civilizations started on Earth a long time ago, notably with the SETI program, but have returned no tangible results so far (see Chapter 6).

### 2.4.2  Searching for habitats

As in our Solar System, the likelihood of finding a planet in the habitable zone of any planetary system depends not only on the presence of such a zone but also on its extent and its distance to the primary star. As we have seen, the planet needs to be inside the circumstellar habitable zone (CHZ), which is limited to a stellar system, whether by the right conditions on its surface or its subsurface.

For example, Ashwini Kumar Lal (2009) estimates that

> a star with 25 % of the luminosity of the Sun will have a CHZ centred at about 0.50 AU. In contrast, a star twice the Sun's luminosity will have a CHZ centred at about 1.4 AU. . . The size and location of CHZs of other stars depend on various factors, such as the size, brightness, and temperature of the star and other planetary factors such as knowledge of a habitable planet's orbit including its eccentricity

It appears, however, that in addition to the CHZ, one must also take into account the conditions in the galaxy in which the planet is located. Indeed, the notion of a galactic habitable zone (GHZ) has emerged, defended essentially by Guillermo Gonzalez in 1995, who imagined a spherical band in which terrestrial life might develop in a galaxy if the right conditions were united: the star has to have the right amount of metals (see Box 5.4), the right orbit, rather circular, around the galactic centre, and the right distance, far enough from the centre so as not to be subject to extreme radiation or gravitational forces damaging to carbon-based life. The galactic centres themselves are often not hospitable environments for planets because of the lack of dust in most cases and the few unstable and old stars that populate them. Planetary migration, as we have described in the previous chapter, sometimes helps to bring all of these requirements together. If a planet is within a CHZ which lies also within a GHZ, then it has a fair chance to develop and sustain lifeforms.

In our Galaxy (the Milky Way), the GHZ is currently believed to be a slowly expanding region as the heavy elements build up. In a recent publication dealing with habitability in the GHZ, Gowanlock, Patton and McConnell (2011) estimate that at least half of the habitable planets should be located preferably towards the inner Galaxy ($R <$ 4.4 kiloparsec), well above the galactic midplane. Depending on the models used, however, the dimensions and location of the GHZ differ. It is even more difficult then to define such parameters for other galaxies, as they vary even more in their compositions, and may have a larger or smaller GHZ – or none at all.

The Drake equation tries to estimate the likelihood of alien intelligent life and needs to include a factor $(n_e)$ for the average number of habitable planets in a planetary system. In 2010 Paul Butler and colleague Steven Vogt (Vogt *et al.*, 2010) of the University of California at Santa Cruz announced the discovery of a planet, Gliese 581G, which is considered to be in the habitable zone of its star because of its distance from its primary and its size, which suggest there could be liquid water on the surface and that the planet is large enough for its gravitational forces to be able to retain an atmosphere around it. Therefore, the basic conditions for life are present to allow it to begin and keep it going (see Chapter 6 for more details). And from current research, it would seem that similar conditions are present in many other Solar Systems, contradicting the Rare Earth hypothesis and pushing our planet even further away from its uniqueness than Copernicus had already.

### 2.4.3 Searching with what?

But what means do astronomers use in their search for habitats? Over the past decades, continuous progress has been achieved in the development of telescopes and instruments. We now benefit from 10-metre class telescopes, which allow us to track fainter and fainter objects and to study their physical properties. As an example, the Kuiper Belt objects have been discovered thanks to ground-based photometric observing campaigns. The quality of images has also significantly improved with the development of a new technique, called adaptive optics. It allows the astronomer to correct the atmospheric turbulence, a limiting factor for the quality of astronomical images. The method consists in observing a reference star close to the object to be measured, and using a small deformable mirror with actuators to correct the image of the star in real time. This method makes it possible for large telescopes to reach the diffraction limit (the smallest angular distance measurable on the sky, which is inversely proportional to the telescope diameter). The use of adaptive optics, combined with 10-metre sized telescopes, has allowed astronomers to obtain high-quality images of planets and satellites, and thus monitor seasonal and climatic variations (a striking example is the big storm which has been developing on Saturn since December 2010). By taking images at different wavelengths, probing different altitude levels, astronomers are able to build 3D images of planetary and satellite atmosphere and to monitor the spatial and temporal variations.

In addition to imaging techniques, spectroscopy is a favoured means for exploring the chemical composition of planetary atmospheres and surfaces. Spectroscopy has been successfully performed from the ground (for instance with ESO's Very Large Telescope in Chile; Figure 2.17). Ground-based spectroscopy has been a powerful tool for identifying the atmospheric composition of the giant planets, and is starting to give promising results for the characterization of exoplanets' atmospheres.

Both ground-based and space observations are needed and are fully complementary for exploring the Solar System and beyond. Astronomical satellites (the Hubble Space Telescope, Figure 2.18, in

FIGURE 2.17  The four 8-m units of the Very Large Telescope with the four 1.8-m auxiliary telescopes composing the VLT Interferometer at Cerro Paranal in Chile. (Image © ESO/H. Heyer.)

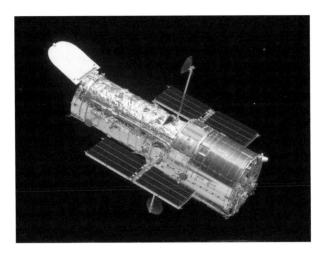

FIGURE 2.18 The Hubble Space Telescope. (Image courtesy of NASA/ ESA/STScI.)

the ultraviolet and visible range, ISO and Spitzer in the infrared range, Herschel in the submillimetre range) have given or are giving key information on the nature of planets and exoplanets, while the CoRoT and Kepler missions have opened the new research fields of transiting exoplanets. Finding Earth-sized habitable planets, for instance, is a key objective of the Kepler mission space observatory, which in March 2009 launched a telescope with a 1.4-metre primary mirror to survey and compile the characteristics of habitable-zone planets. As of December 2012, Kepler has discovered over 2000 candidate exoplanets, with several tens of those candidates located within the Goldilocks zone (see Subsection 5.1.4). One of these, recently announced by the name KOI-172.02, is thought to be an Earth-sized exoplanet orbiting around a G-type star in the habitable zone. Awaiting confirmation, these results still show a trend for the existence of several planetary systems with the right habitable conditions in our Galaxy.

Back to the Solar System planets, *in-situ* space exploration has provided us with a wealth of discoveries, ranging from the first images of the surfaces of Mars, Venus and Titan, to the chemical analysis of their atmospheres. In the case of Mars, important milestones were, in the 1970s, the Mariner 9 orbiter and the two Viking missions, which both included an orbiter and a lander (see Subsection 3.3.2). The exploration of Venus benefited from the Venera spacecraft in the 1980s, and the Magellan radar mission which mapped its surface in the early 1990s (see Section 3.2). Our understanding of terrestrial and outer planets has been completely renewed by space exploration, in particular the Galileo mission which explored the Jovian system between 1995 and 2003, and the Cassini mission, in operation in the Saturn system since 2004. There is no doubt that both ground-based and space observing campaigns will continue together to feed our search for habitable worlds in the Universe. Our search will also benefit from the continuous developments of theoretical modelling, numerical simulations and laboratory simulation experiments.

In what follows, we will adopt the definition of habitable conditions (not to be confused with the presence of life) as resting on the fulfilment of four conditions: water, elements, energy and time. The search for habitats other than our own planet in the Solar System and beyond is of course conditioned by our human need to comprehend how our own planet became habitable, and to see if there is reason to believe that other such environments exist. In the case of a positive answer to the latter question, we could pursue the exploration further by looking for life or other forms of life.

As we have seen, the most crucial factor in the evolution of terrestrial life has been the availability of liquid water. Not only does water serve as the solvent for biochemical reactions but it also furnishes the hydrogen upon which living matter depends. Water will remain liquid under a pressure of 1 bar (terrestrial sea-level pressure) between 0 °C and 100 °C. But life has proved itself tough here on Earth, and it could perhaps thrive in more extreme environments, as we have seen in the previous section. And of course, it might be that life can develop along very varied lines. If it could be based on ammonia as a solvent, instead of water, this would allow it to subsist at low temperatures. Then, again, it may be that life is possible in the atmospheres of Jupiter-like planets or in the undersurface oceans of icy moons. Such alternative biological forms would extend the habitable zones of stars beyond those considered traditionally. The boundaries of habitable zones have been evolving, and our search for hospitable environments in the Solar System and beyond is now even more challenging and exciting.

This is what we will investigate in the following chapters.

# 3    Terrestrial planets and their diverging evolutions

In the original definition of the habitable zone, its boundaries first encompassed the orbits of Venus to Mars, planets close enough to the Sun for solar energy to drive the chemistry of life – but not so close as to boil off water or break down the organic molecules on which life depends. These planets and their neighbourhood have been observed from the ground since the earliest times and explored *in situ* since the beginning of the space age. What have we found so far?

Although they are all members of the terrestrial planet family, Mercury, Venus, the Earth and Mars have very distinct properties. These differences are basically the result of two factors: their heliocentric distance (and hence their temperature) and their mass. Mercury, Venus and Mars, all easily visible with the naked eye, have been known since antiquity and celebrated in every mythology. But their analysis as physical objects has mostly developed since the beginning of the twentieth century, with the advent of spectroscopy and photography. A new era started in the 1960s with the advent of space exploration. The space adventure, marked with many failures, continues today. Two orbiters, Messenger and Venus Express, are currently observing Mercury and Venus, respectively; several orbiters are operating around Mars and a sophisticated rover, Curiosity, is exploring the nature of the Martian surface.

## 3.1    LOOKING OUT FROM MERCURY'S DESERT

Mercury, the planet closest to the Sun, is also the smallest planet. Located at 0.4 AU from the Sun, the planet is in slow rotation (it takes 58 terrestrial days to rotate on its axis (called a sidereal day), and 88 days to orbit around the Sun). The Mercurian day is thus very long (sunrise to sunrise it is 176 terrestrial days) which translates into very strong temperature contrasts between day and night. The contrast is amplified by the absence of atmosphere: owing to the low gravity and the high dayside temperature, the planet cannot retain a stable atmosphere. The surface temperature ranges from about 700 K on the dayside down to 90 K on the nightside.

The surface of Mercury is heavily cratered and shows many similarities with the lunar surface. Its exploration started in the 1970s with the Mariner 10 mission which achieved several flybys of the planet and mapped most of its surface. Ground-based optical and radar observations also contributed to our knowledge of the surface and showed evidence for a transient exosphere, resulting from the interaction of the solar wind with the surface. A surprise result was the detection by Mariner 10 of a strong magnetic field, unexpected on such a small object; this might indicate the presence of an iron-rich core surrounded by a fluid conductive envelope. Another unexpected discovery, from the radar observations, was the possible presence of water at the poles of Mercury, in some localized regions in permanent shadow. Water could come from the interaction with the solar wind or from meteoritic bombardment or outgassing. Since 2008, Mercury has again been explored by a National Aeronautics and Space Administration (NASA) space mission, Messenger (Figures 3.1, 3.2), which is at present in orbit around the planet after three successive flybys. Another mission devoted to Mercury, Bepi–Colombo, will be launched by the European Space Agency (ESA) in 2014 for an in-depth exploration of the planet in 2020 and beyond.

From the point of view of exobiology – our search for life beyond our own planet – Mercury is of little interest. The absence of stable atmosphere and the extreme temperatures cannot have been favourable for the emergence of life at any time of its history. Even if traces of permanent water-ice may be present in the polar regions, conditions for maintaining liquid water have never been met. However, as an end-member of the rocky planet family, its exploration is worthwhile from a physical point of view, in particular to understand the origin of its magnetic field, to study its magnetosphere and its transient atmosphere, and to solve the enigma of the possible polar water reservoirs. This exploration takes on a new significance in view of the

FIGURE 3.1 Planet Mercury as seen by the Messenger spacecraft, in orbit around the planet since March 2011. The surface of Mercury is heavily cratered and very similar to the lunar surface. (Image courtesy of NASA NSSDC.)

recent discovery of small extrasolar planets in the close vicinity of their stars.

The three other terrestrial planets, Venus, the Earth and Mars, share two properties of major interest for exobiology: first, they all have a stable atmosphere; second, water is present in that atmosphere (Figure 3.3). In the case of Venus and Mars, liquid water is absent today and water vapour is only a trace component; however, as we will see below, there is good evidence that liquid water was present in abundance in the past history of Mars, and it is reasonable to consider that it may also have been the case on Venus. These two planets are thus favoured targets in our search for past or present habitats.

## 3.2 A PAST OCEAN ON VENUS?

Venus, known since antiquity as the 'shepherd's star', is our closest neighbouring planet and also the most similar one to Earth, in terms of

FIGURE 3.2  Fractured terrain on Mercury: At the antipode of Caloris Basin, the so-called 'Weird Terrain' is a region of unusual, hilly landscape. It might have been generated by shock waves produced during the Caloris impact and converging at the opposite point. Another possible origin is the convergence of ejecta. (Image courtesy of NASA/JPL/Northwestern University.)

FIGURE 3.3  The three terrestrial planets with an atmosphere, showing their relative sizes. Mars (left) is the smallest; its surface is covered with deserts and the polar caps are visible; the Earth (centre) shows a large-scale cloud system of water-ice; Venus (right) is covered with a thick cloud of sulfuric acid. For colour version, see plates section. (Image courtesy of NASA.)

size, mass and radius. Located at 0.7 AU from the Sun, Venus receives more solar energy than the Earth. Its surface temperature is thus expected to be higher than on our own planet. In fact, the temperature at the surface (730 K) of Venus is by far higher than expected, owing to a huge greenhouse effect as will be discussed below. The atmospheric composition is dominated by carbon dioxide (96%) with a few per cent of molecular nitrogen, and the surface pressure is 90 bars.

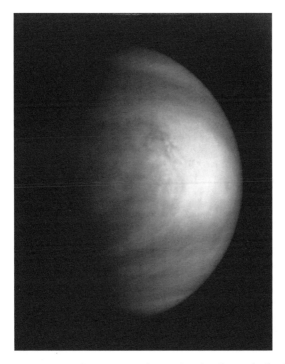

FIGURE 3.4  Planet Venus, observed in the ultraviolet by the camera of the Galileo probe during its flyby of the planet in February 1990. The UV radiation probes the top of the thick cloud layer, mostly composed of sulfuric acid. For colour version, see plates section. (Image courtesy of Galileo Project/JPL/NASA.)

The surface of Venus is hidden below a thick cloud blanket, mostly composed of sulfuric acid (Figure 3.4). It appears, then, that Venus' surface is a particularly inhospitable environment (see Box 3.1).

Little was known about Venus before the second half of the twentieth century. There was no information about the nature of the troposphere below the clouds, and, from the seventeenth century onwards, philosophers and scientists such as Cyrano de Bergerac and Fontenelle dreamed about possible inhabitants living on the Moon, Mars or Venus. The exploration of Venus started in the 1970s, with the first ground-based radio observations of the surface and later with space missions: the Venera spacecraft launched by the Soviet

BOX 3.1   **The sulfur cycle on Venus**

The atmosphere of Venus is characterized by very high temperatures and pressures at the surface (730 K, or 457 °C, and 90 bars). It is mainly composed of carbon dioxide $CO_2$ (96 per cent), $N_2$ (3 per cent), and traces of minor species: $SO_2$, $H_2O$, CO, OCS and $H_2S$ below the per cent level. At altitudes ranging between 50 and 70 km, several cloud layers, mostly composed of $H_2SO_4$, hide the surface at visible wavelengths.

The sulfur cycle is known to play a key role in the atmospheric chemistry of Venus. Below the clouds, the main minor species is $SO_2$ (100 ppm), possibly of volcanic origin, then $H_2O$ (30 ppm), then CO and OCS (about 10 ppm) and then $H_2S$ (about 1 ppm). The species are transported up to the cloud level by convection. Above the clouds, all species are photodissociated by the solar ultraviolet radiation: $H_2SO_4$ is

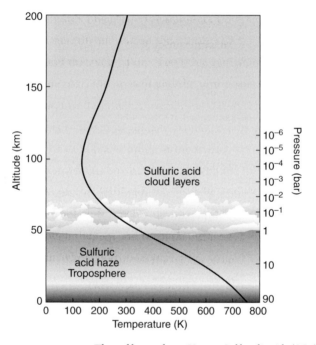

FIGURE BOX 3.1   The sulfur cycle on Venus. Sulfur dioxide ($SO_2$) is abundant below the clouds, and reacts at the cloud level with $H_2O$ to form the $H_2SO_4$ cloud. (Adapted from the *Encyclopedia of Science*, http://www.daviddarling.info/encyclopedia/V/Venusatmos.html)

BOX 3.1 **(cont.)**

dissociated into $SO_3$ and $H_2O$, then into $SO_2$, SO and possibly other sulfur components. The $SO_2$ is also dissociated over a very short timescale (less than a day). Above the clouds, $SO_2$ reacts with atomic oxygen to form $SO_3$ which combines with $H_2O$ to reform $H_2SO_4$, whose condensation feeds the cloud layers. Subject to evaporation in the lower levels and condensation at the top, the $H_2SO_4$ cloud is thus constantly dissociated and replenished by the $SO_2$ cycle.

Above the clouds, all gaseous species (except CO) are strongly depleted with respect to their tropospheric abundances, in particular $SO_2$ and $H_2O$. Higher in the mesosphere, at about 90 km, there seems to be evidence for an increased abundance of $SO_2$ whose origin is still unclear. It might imply the existence of a second reservoir of sulfur, possibly in the form of $S_x$.

Union in the 1970s and 1980s, the Pioneer Venus (1978) and Magellan (1991) NASA missions, and, more recently, the European Venus Express mission (2006) still in orbit around the planet. In the 1960s, the high surface temperature was discovered from radio measurements, while the presence of massive amounts of $CO_2$ was inferred from infrared spectroscopy. First images of the surface were revealed by the Venera 13 lander in 1983 (Figure 3.5). Radar measurements, first from the ground and later from Magellan, revealed the surface topography to be dominated by volcanoes (Figure 3.6). Crater counting over these features has shown that the surface of Venus is remarkably young, less than a billion years old. Unfortunately, there is no way of finding traces of the early history of Venus, as most of the traces of the planet's history have been wiped out by surface remodelling.

Can we find early diagnostics by studying the present atmosphere? The answer is yes, through the study of deuterated water, or 'heavy water'. Water ($H_2O$) is present in different forms, the usual one with the formula $H_2^{16}O$ and its isotopic species, among which are HDO (with one deuterium, D, atom) and $H_2^{18}O$ (see Section 1.2). In

FIGURE 3.5 The surface of Venus as first revealed by the Venera 13 mission in March 1982 (Academy of Science of the USSR. Image courtesy of NSSDC/NASA.)

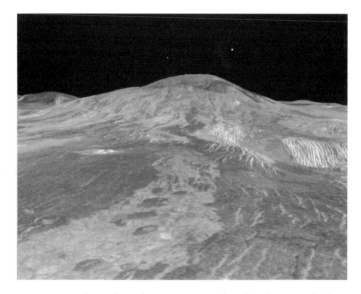

FIGURE 3.6 The surface of Venus is covered with volcanoes. The low number of craters indicates that these volcanoes are relatively young (less than one billion years). Volcanism on Venus might still be active but there is still no evidence for it. The image is a reconstruction from the radar images taken by the Magellan spacecraft in 1991–1992. It shows a perspective view of Sif Mons, a mountain 2 km high and 300 km in diameter. The view is looking south from 360 km north of Sif Mons at an elevation of 7.5 km. This image has a vertical exaggeration of a factor of 23. (Image courtesy of NSSDC/NASA/JPL.)

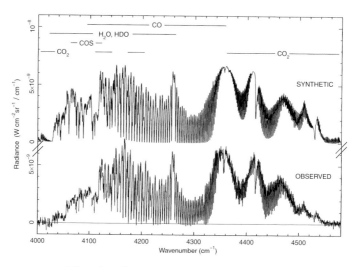

FIGURE 3.7  The infrared spectrum of Venus observed with the Fourier Transform Spectrometer of the CFH (Canada–France–Hawaii) telescope in 1990. Top: model; bottom: observations. Comparison of absorptions due to $H_2O$ and HDO is used to infer the D/H ratio in the atmosphere of Venus. (From Bézard *et al.*, 1990.)

today's atmosphere, water vapour is a very minor constituent: only 30 ppm (i.e. part per million) under the clouds, and even less above, as water reacts with sulfur dioxide to form $H_2SO_4$, the main cloud component. Heavy water HDO is even much less abundant. On Earth, the D/H ratio, as measured from the $HDO/H_2O$ ratio in the oceans (the Vienna Standard Mean Ocean Water – or VSMOW – value), is $1.6 \times 10^{-4}$ (see Section 1.2). If Venus and the Earth had evolved in the same way, we would expect D/H to be equal on both planets. But that is not the case at all: the D/H ratio on Venus, as measured from $H_2O$ and HDO infrared spectroscopic measurements (Figure 3.7), is about 120 times the terrestrial value in the lower troposphere (below the clouds), and more than 200 above the clouds. What could be the origin of such an excess? The most likely explanation is that water was very abundant in the early history of Venus, possibly as abundant as on the Earth today. At the time of the planets' formation, the Sun was fainter than today by about 30 per cent in flux. This slowly increased as the Sun followed its stellar evolution, leading to an increase in the temperature. As a result, water escaped continuously, but with a differential effect, the heavy water HDO escaping more slowly than the normal water $H_2O$. As a result, the D/H value increased with time up to the high value observed today.

At the time of Venus' formation, the effective temperature of the planet (corresponding to the absorbed fraction of the solar flux received at that time), in the absence of a greenhouse effect, was probably in the range 270–350 K, appropriate for water to be in liquid form. We can thus speculate that Venus did host a liquid ocean of water in the early phases of its history. As the solar flux and thus the atmospheric temperature increased, the water vaporized, increasing the greenhouse effect, already fed by the large amounts of carbon dioxide. As a result, a runaway greenhouse effect took place, leading to the very high surface temperature now observed at the surface of Venus. Water vapour was dissociated by the solar ultraviolet flux; hydrogen atoms were light enough to escape. The heavier oxygen atoms also escaped, in a mechanism which is not completely understood today.

Could life have appeared on Venus in the first billion years, at the time of the water ocean? From what we know of the evolution of solar-type stars, the hypothesis is not unreasonable. However, we will probably never have the answer to this question, as the recent surface remodelling has erased any signatures of the planet's early days.

## 3.3   LIFE ON MARS? AN OLD QUEST AND A MODERN CHALLENGE

For centuries, Mars, the red planet named after the God of War, has been the focus of human dreams of possible life beyond Earth. Its proximity and relative similarity to the Earth, with a thin atmosphere, and a surface visible from Earth with deserts, volcanoes and icy polar caps, made planet Mars a favoured target for hosting extraterrestrial life. Past and present literature (such as the famous *War of the Worlds*, written by H. G. Wells in 1898), is filled with stories of 'Martians', or 'little green men' poised for invading the Earth. From fiction to science, Mars has been and continues to be a prime candidate in the search for habitable worlds, and this quest is now developing in the framework of an ambitious space exploration programme, with, in some minds, the prospect – still controversial – of a future manned exploration of Mars (see Chapter 6).

### 3.3.1  Schiaparelli's canali

In 1877, these dreams took on a scientific dimension for the first time when the Italian astronomer Giovanni Schiaparelli announced the discovery of a network of linear features ('canali') on the surface of the planet. The astronomer was cautious about their interpretation, but others were less so. These features were searched for by other observers, confirmed by some and ruled out by others, raising controversy within the scientific community. Camille Flammarion, in his book about Mars and its habitability, published in 1894, developed the first theory of possible life on Mars. The American astronomer Percival Lowell was the first to interpret the 'canali' as evidence for an intelligent civilization building irrigation channels (it must be recalled that, at that time, Mars was already known to be a very dry planet). Percival Lowell actually devoted his life to this research, building his own observatory for this purpose in the early 1900s. He reported the detection of hundreds of canali, some of them with a double structure interpreted as built on purpose by the 'Martians' to prevent possible failures (Figure 3.8).

This idea, however, was received with scepticism by other astronomers, in particular the French astronomer Eugene Antoniadi. Initially a supporter of the canali interpretation, he carried out an observing campaign over several years, using the big refracting telescope at Meudon Observatory and then a new telescope at Pic du Midi Observatory. In 1909, he came to the conclusion that the so-called 'channels' were no more than an optical illusion (Figure 3.9). His results were published in 1930 in his book *The Planet Mars* (in translation: Antoniadi, 1975).

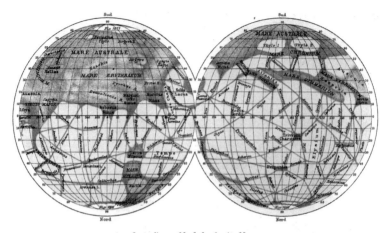

Carte d'ensemble de la planète Mars
avec ses lignes sombres non doublées
observées pendant les six oppositions de 1877-1888
par J.V. Schiaparelli.

FIGURE 3.8 The surface of Mars as drawn by the Italian astronomer Giovanni Schiaparelli in 1877. The linear features were erroneously interpreted by some scientists as the signature of artificial irrigation channels, and hence the proof of an intelligent life. (Image courtesy of Wikimedia Commons, http://commons.wikimedia.org/wiki/File: Mars_Atlas_by_Giovanni_Schiaparelli_1888.jpg.)

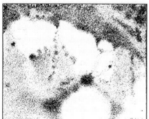

FIGURE 3.9 A comparison of the 'canali' or 'channels' on Mars, as drawn by Schiaparelli in 1877 (top) and later by Eugene Antoniadi in 1930. The images made by Antoniadi at the big refractor of Meudon Observatory demonstrated that the linear features reported by Schiaparelli were only an optical illusion, due to the moderate quality of his instrument. (Image courtesy of the *Encyclopedia of Science*, http:// www.daviddarling.info/encyclopedia/A/ Antoniadi.html.)

For most of the astronomical community, the story was over; but the myth lasted for decades, especially in the United States where the Martian maps used by NASA in the 1960s were still the maps by Lowell showing the canali. The myth actually survived until 1965, when the Mariner 4 space mission revealed the true nature of the Martian surface, covered with craters and devoid of any artificial feature.

### 3.3.2 The Viking mission, or the search for life

Mars, the most remote terrestrial planet from the Sun, is much colder than Venus and the Earth, with temperatures ranging from 150 K to 300 K depending on the location and season, but also much less massive (a tenth of the Earth's mass). Another major difference is its extremely thin atmosphere, with a mean surface pressure of 6 millibars. The atmospheric composition, in contrast, is similar to that of Venus, with 95 per cent $CO_2$, 3 per cent $N_2$ and mere traces of water vapour and CO. As a result of the low temperature, $CO_2$ condenses alternately from one pole to the other following the seasonal cycle, leading to the formation of seasonal polar caps. The surface of Mars, easily observable from the Earth, shows many volcanic and tectonic features, in particular the very high Tharsis volcanoes (including Olympus Mons, at 25 km the highest known volcano in the Universe) and a huge canyon, the Valles Marineris.

The question of a possible form of life, past or present, was still open at the beginning of the space era and, 50 years later, is still with us. In 1965, Mariner 4 revealed a dry planet with no sign of life, but with evidence for volcanism, past tectonics and meteorological activity. After Mariner 9, the search for life on Mars concentrated on very simple bacteria-type organisms, able to survive the harsh environment of the Martian surface in terms of UV solar radiation and cosmic rays. After the first success of the Mariner 9 orbiter in 1972, the search for living microorganisms was the driver behind the ambitious Viking space mission, launched by NASA in 1975. The Viking programme included two orbiters and two landers which monitored the Martian atmosphere and surface for several years (Figure 3.10). It was a

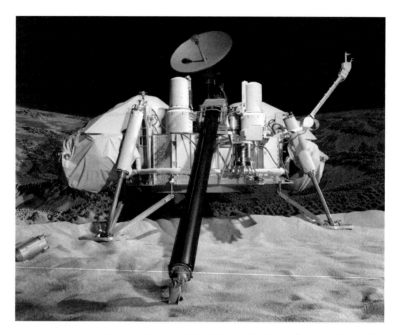

FIGURE 3.10  An artist's view of the Viking 1 lander. It shows at the top the antenna for Earth communication and, on the front, the robotic arm designed for collecting samples. (Image courtesy of NASA/JPL-Caltech/ University of Arizona.)

tremendous scientific and technical success, and led to a reference database still in use today.

Among the lander experiments, four instruments were designed to search for signs of biological activity: (1) the Gas Chromatograph/ Mass Spectrometer (GCMS) for detecting organic molecules, (2) the Gas Exchange (GE) experiment for detecting the outgassing of a sample fed with nutrients, (3) the Labeled Release (LR) experiment for detecting an excess of $^{14}CO_2$ from a sample fed with water and nutrients, and (4) the Pyrolytic Release (PR) experiment for detecting possible photosynthetic processes. The surface samples were collected by a robotic arm and carried into containers on the spacecraft for analysis.

While the results of the other experiments were negative, the interpretation of the Labeled Release experiment raised strong controversy in the community. Its objective was the search for hetero-trophic organisms (i.e. those fed with organic matter) through the detection of a $^{14}CO_2$ excess. The LR experiment did detect an excess of $^{14}CO_2$ when the Martian soil was first exposed to a medium enriched in water and nutrients. However, this result contradicted the negative results of the other three experiments, in particular the total absence of organic molecules revealed by the gas chromatograph and mass spectrometer. The final consensus in the community was that the effect measured by the LR experiment was probably the result of purely chemical processes, and that there was no evidence for bio-logical activity at the surface of Mars. The absence of organic mol-ecules at the surface was attributed to the strong UV solar flux which reaches the surface of Mars, and to the presence of a strong oxidizer, possibly hydrogen peroxide $H_2O_2$. This molecule was actually detected at trace abundances (about $10^{-8}$) in the Martian atmosphere in 2004. Another possible oxidizer is perchlorate, detected in 2008 by the Phoenix lander; it was suggested that its presence might have been responsible for the false positive LR detection.

Still, the consensus about the absence of biological activity on Mars is not shared by all. In particular, Gilbert Levin, Principal Investigator of the LR experiment, has claimed that LR was more sensitive than the other experiments, in particular the GCMS which might have missed the detection of organic molecules. It has also been suggested that the Viking experiments may not have been sen-sitive enough to detect traces of organics, or that the technique of heating samples may have led to the destruction of organics, if present, by perchlorate. In summary, the debate is not over, and more information should come from the exploration of the Martian soil by Mars Science Laboratory/Curiosity, in operation since August 2012 (see Section 3.3.3 below).

### 3.3.3 'Follow the water!'

In spite of its spectacular scientific achievement, the Viking mission, as seen from an astrobiological point of view, was considered as a failure by the community at large. The direct consequence was, in the United States, an interruption to the Martin space programme which lasted for more than 20 years. But speculation about the possible existence of past or present traces of life did not disappear. It became clear that the possible 'niches' had to be sought in regions permanently protected from sunlight, or at some depth below the surface. In addition, a surprising discovery, made by Mariner 9 and then the Viking orbiters, was the identification of several types of valley network, some in very old terrains, others (called 'outflow channels') in regions disturbed by catastrophic events (chaotic regions). These images led to the first evidence that liquid water was present on Mars at some epochs of its history. As a result, when the search for life on Mars started again in the 1990s, the main driver was 'Follow the water!' In other words, look for places where liquid water may have flowed in the past. Twenty years later, the objective is still the same, but significant progress has been achieved in our understanding of the history of water on Mars (Figure 3.11).

In spite of an impressive record of failures, several space missions, mostly from NASA – orbiters, landers and rovers – have successfully contributed to our knowledge of the Martian atmosphere and surface. In particular, the Mars Global Surveyor orbiter, in 1998, indirectly detected the presence of a remnant crustal magnetic field, a diagnostic of the existence of a dynamo in the first billion years of the planet's history. The extinction of the dynamo and the disappearance of the magnetosphere may have favoured the outgassing of the Martian atmosphere and the low surface pressure observed today. The gamma-ray experiment on Mars Odyssey, in 2000, detected the presence of hydrogen atoms (mostly likely due to the presence of $H_2O$) under the poles of the planet (Figure 3.12). More recently, the infrared imaging spectrometer OMEGA on the Mars Express orbiter, launched by the European Space Agency ESA in 2003, detected the presence of phyllosilicates (or clays) in the oldest terrains of the southern hemisphere, and found evidence for sulfates in other more chaotic areas. The presence of clays was

FIGURE 3.11 Mosaic of the Valles Marineris hemisphere of Mars projected into point perspective, a view similar to that which one would see from a spacecraft. The viewer's distance is 2500 km from the surface of the planet. The mosaic is composed of 100 Viking Orbiter images of Mars. The centre of the scene (latitude −7°, longitude 78°) shows the entire Valles Marineris canyon system, over 3000 m long and up to 8 km deep, extending from Noctis Labyrinthus, to the west, to the chaotic terrain to the east. The three Tharsis volcanoes (dark spots), each about 25 km high, are visible to the west. For colour version, see plates section. (Image courtesy of NASA/MDIM.)

interpreted as the signature of abundant and permanent liquid water in the early phases of Martian history (with flooding by this water being responsible for the valley networks; Figure 3.13), while the presence of sulfates was associated with the temporary catastrophic flow of water in the more recent past, responsible for the outflow channels mentioned above. These results led to new insights in the quest for past forms of life: they have been and

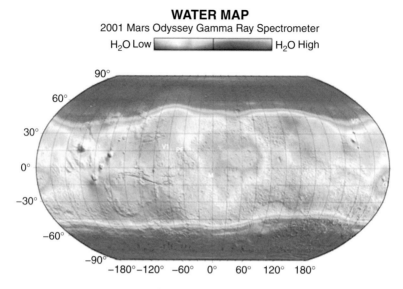

**WATER MAP**
2001 Mars Odyssey Gamma Ray Spectrometer
H₂O Low $\quad$ H₂O High

FIGURE 3.12 Water under the surface of Mars, as detected by the Gamma Ray Spectrometer of the Odyssey mission in 2000. The instrument measured the nuclear radiation emitted by the surface, following the impact of cosmic rays. The abundance of neutrons emitted by the surface is an indicator of the content of hydrogen atoms below the surface. A low content of neutrons indicates high hydrogen abundance and thus a high water content. In each hemisphere, the map was obtained during summer, to avoid the presence of the carbon dioxide seasonal cap. For colour version, see plates section. (Image courtesy of NASA/JPL.)

will be used to select the most appropriate sites for future *in situ* exploration.

Among several important scientific achievements, the Mars Global Surveyor (MGS) spacecraft, with its laser altimeter, made a precise measurement of Martian altitudes, and this has potential implications for the history of water on Mars (Figure 3.14). The Martian surface is known to be made of two different hemispheres: the northern part is made of low-altitude, relatively recent plains, while the southern part is made of older highlands (this is where the fossilized crustal magnetic field was detected). Several explanations have been proposed to account for this difference, some claiming it is of

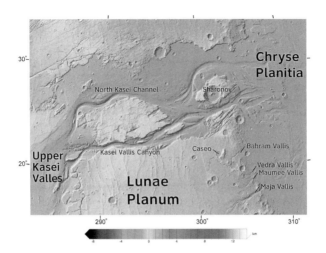

FIGURE 3.13  An example of a valley network (Kasei Valles) on Mars, near Chryse Planitia, indicating the presence of liquid water in the past history of the planet. For colour version, see plates section. (Image courtesy of NASA/Google Earth, via Wikimedia Commons.)

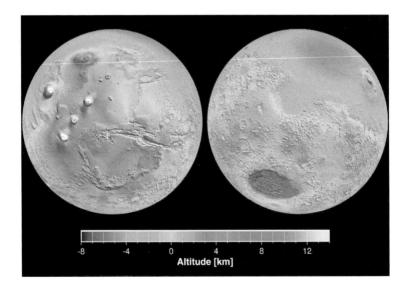

FIGURE 3.14  Altimetry (height mapping) of Mars, as measured by the laser altimeter experiment (MOLA) of the Mars Global Surveyor mission. The white dots over the red area on the left image are the Tharsis volcanoes. The dark blue spot on the right image is the bottom of Hellas basin, a giant impact crater. For colour version, see plates section. (Image courtesy of NASA/JPL.)

FIGURE 3.15  A self-portrait of Curiosity, the rover of the Mars Science Laboratory mission, from a combination of several high-resolution images taken by the Mars Hand Lens Imager. The rover has been in operation on the surface of Mars since August 2012, when it landed in Crater Gale. The robotic arm is designed to collect samples at the surface and below. The samples are analysed by the rover instruments for chemical and mineralogical characterization. For colour version, see plates section. (Image courtesy of NASA/JPL-Caltech/Malin Space Science Systems.)

internal origin (through convection), others favouring an external origin (from impacts), but no firm conclusion has been reached today. The two regions are separated by the so-called 'dichotomy line'. At the level of this dichotomy line, the MGS laser altimeter found a line of remarkably constant altitude over a thousand kilometres, which suggests it might have been the shoreline of a previous boreal ocean of liquid water. This assumption raised objections, in particular because of the absence of detection of sedimentary materials in this region. However, in 2012, the Mars Express radar experiment (MARSIS) measured a low dielectric constant (implying the presence of sedimentary or aqueous materials below the surface) in a region corresponding exactly to the one indicated by the MGS altimeter results. The question of a past boreal ocean on Mars is thus still open.

Another insight into the history of water on Mars was provided over 20 years ago by the measurement of the deuterium abundance in today's Martian atmosphere. Using near-infrared spectroscopic measurements of HDO and $H_2O$, just as on Venus, a D/H measurement was inferred, about five times larger than the terrestrial value. Although much less than in the case of Venus, this significant excess was attributed to the differential escape of water vapour over the planet's history (see Sections 1.2.3 and 3.2). The excess was smaller than on Venus because the early atmosphere, although more dense than today, was still much more tenuous than the primordial atmosphere of Venus. Another isotopic measurement, $^{15}N/^{14}N$, obtained by the mass spectrometers of the Viking probes in 1976, led to a consistent conclusion: the surface pressure in the early atmosphere of Mars might have been as much as 0.1 atm, i.e. more than ten times its present value.

A new milestone has been reached with the launch of the Mars Science Laboratory mission (now called Curiosity) in November 2011. This sophisticated vehicle landed on Mars on 5 August 2012, with an impressive set of instrumentation for *in situ* analysis of the Martian soil. Five times heavier than its predecessors, Spirit and Opportunity, launched by NASA in 2003, the payload of Curiosity includes in particular cameras, a laser-induced breakdown spectrometer, a high-energy spectrometer, a mass spectrometer and a tunable laser spectrometer for searching for organics in both the gas and solid phases. Among its first results, Curiosity has reported the absence of detectable methane at the surface of Mars (see Box 3.2). On the European side, the ExoMars mission, in cooperation between ESA and Russia, includes an orbiter, to be launched in 2016, for an in-depth exploration of the Martian atmospheric composition, and a rover, to be

BOX 3.2 **Methane on Mars?**

In 2004, the astronomical community was agitated by a debate with important potential implications for exobiology. Within about a year, several teams reported the identification of traces of methane in the atmosphere of Mars. The spectroscopic data came from different origins, including ground-based telescopes but also space data, in particular the PFS instrument aboard Mars Express, and the TES instrument aboard the Mars Global Surveyor. Reported abundances ranged between 0 and about 40 ppb.

If confirmed, the presence of methane on Mars could have different origins, including a biogenic one: on Earth, most of the atmospheric methane is of biogenic origin. But the methane could have a geological source. Volcanism seems to be excluded, as a very low upper limit (0.3 ppb) has been found for the $SO_2$ gaseous content, which would be expected to be more abundant than methane by orders of magnitude. Another geothermal mechanism involves reactions between water and rock, and the formation of $H_2$, then $CH_4$ and hydrocarbons through Fisher–Tropsch reactions and the so-called serpentinization process. Another possible source would be the sporadic outgassing of underground pockets of clathrate hydrates – ice-like crystalline solids with methane trapped within a hydrogen-bonded lattice.

A comparison of the various data sets soon demonstrated that the methane emission, if real, had to be localized in time and space. It also appeared that several reported detections were at the noise limit and were thus marginal. In terms of data quality, the most convincing observations were made by Michael Mumma and his colleagues in the spring of 2003. Spectral images showed evidence for two spectroscopic transitions characteristic of methane. Another more marginal detection was reported by the same team in 2006. However, all subsequent observations by the same team and by others have failed to detect methane again.

These observations raised a real problem for atmospheric physicists. It could be imagined that a single outgassing event of methane took place in 2003. But to explain its absence detection in subsequent observations, the methane must have had an extremely short lifetime.

BOX 3.2 **(cont.)**

The photochemical destruction lifetime of methane on Mars is about 350 years. No mechanism has been found to explain such a fast destruction rate.

The presence of methane on Mars has been questioned by some authors, including Kevin Zahnle and his colleagues, who argued that there might have been a misinterpretation of the Martian data due to poor correction of the terrestrial isotopic lines of methane. The debate is still going on.

A new piece of information has been delivered by the Curiosity rover in November 2012. A first analysis of the Tunable Laser Spectrometer experiment reported the absence of methane at the 5 ppb level. More results are expected from Curiosity in the coming year.

launched in 2018, which will include a drill for extracting soil and analysing samples from the deep subsurface. In the longer term, the main objective of the space exploration of Mars will be the Martian Sample Return mission. This will make it possible to carry out the same measurements on abundances of elements and isotopes, and thus dating measurements, on Martian samples as achieved with great success on lunar samples and meteorites.

### 3.3.4 The mystery of ALH84001

There is another way of exploring the soil of Mars to decipher its origin and evolution: it consists in analysing meteorites coming from the planet and collected on Earth. Such rare bodies exist: about a hundred samples among the tens of thousands of meteorites reaching the Earth every year come from Mars. Their Martian origin is testified by their elemental and isotopic compositions, which are exactly similar to those measured by the Viking landers in 1976, for both rocks and gases.

There are three main types of Martian meteorites known together as 'SNC meteorites': the Shergottites (named after the

FIGURE 3.16 The meteorite ALH84001 was collected in Allan Hills, Antarctica, in December 1994. It was identified as a Martian meteorite from elemental and isotopic analysis of its rock and gas components. In 1996, NASA announced that the meteorite could exhibit a signature of past fossil life, but the subject is a matter of great controversy. (Image courtesy of NASA/JSC.)

Shergotty meteorite, found in India in 1865), the Nahklites (after the Nahkla meteorite, which fell in Egypt in 1911) and the Chassignites (after the Chassigny meteorite, which fell in France in 1815). The age of Shergottites appears to be surprisingly young (180 million years) in view of the mean age of the Martian surface. The Nahklites, like the Shergottites, are igneous rocks, but much older ones (1.3 billion years), which were ejected from Mars about ten million years ago and fell on Earth within the past ten thousand years. Like the Nahklites, the Chassignites, about 1.3 billion years in age, were ejected from Mars about 10 million years ago.

Apart from the SNC types, a few other Martian meteorites have been found. The most famous of them is known as Allan Hills 84001, more commonly called ALH84001 (Figure 3.16), found in Antarctica in 1984. This meteorite received much attention from the exobiology community, because a group of scientists identified some structures, revealed by the electron microscope, as fossilized remnants of nano-bacteria and thus as evidence for life. This interpretation, however, was questioned by others who suggested that the observed features could result from terrestrial contamination. A long controversy developed within the scientific community, and there is still a vivid discussion about the biological or non-biological origin of the observed features (see Box 3.3).

BOX 3.3 **Details of the mystery of ALH84001**

The ALH84001 meteorite was discovered on 27 December 1984 in Allan Hills, Antarctica. It is an achondrite (i.e. stony) meteorite, made of low-Ca orthopyroxene, with a mass of 1.93 kg. From its elemental and isotopic composition, identical to that of the Viking samples, it was soon identified as a Martian meteorite, but different from the SNC types. Its history was traced from the use of radiometric dating techniques with several elements of different lifetimes (the decay of rubidium to form strontium (Rb–Sr), potassium to argon (K–Ar), and $^{14}C$ to the stable form $^{12}C$). ALH84001 originated some 4.1–3.9 billion years ago, at a time when liquid water may have been present on Mars. It was blasted off the planet some 15 million years ago and arrived on Earth about 13 000 years ago.

In August 1996, ALH84001 became famous all over the world when David McKay and his team reported the presence of traces of life in the sample. The news, announced publicly by NASA and even the President of the United States, caused enormous interest and emotion, but also

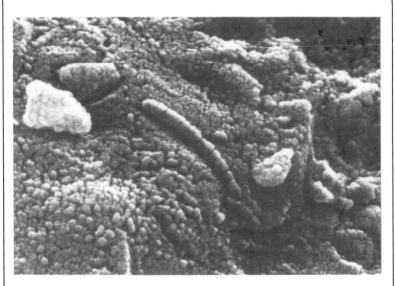

FIGURE BOX 3.3 The structures of ALH84001 observed with a scanning electron microscope (Image courtesy of NASA/JSC, via Wikimedia Commons.)

BOX 3.3 **(cont.)**

considerable controversy within the scientific community, and the debate is still going on today.

What are the original data? Using the scanning electron microscope, McKay reported the presence of structures, 20 to 100 nanometres in diameter, which were interpreted as theoretical nanobacteria and fossilized forms of living organisms. However, no cellular life of this size has ever been observed on Earth; it was also pointed out that nanobacteria might be too small to contain RNA. Another comment was that the biomorphs observed in the meteorite might not originate from Mars, but might be the result of terrestrial contamination. In addition, amino acids and polycyclic aromatic hydrocarbons (PAHs) have been also found in the meteorite. However, there is no evidence for an exobiological origin (PAHs, in particular, are found in many astrophysical environments, such as the interstellar medium and circumstellar envelopes). A plausible explanation for carbonate globules on Mars is formation at high temperature by volcanism or impact, in the absence of any biological processes. In November 2009, McKay and his team reasserted their conclusions about the likely biogenic origin of the ALH84001 features; however, the debate is still ongoing.

## 3.4 BETWEEN VENUS AND MARS, THE EARTH...

In spite of their very different physical conditions, in terms of surface pressure and temperature, the atmospheres of Venus and Mars have a remarkable common property: their chemical compositions are similar. Both atmospheres are largely dominated by carbon dioxide with a few per cent molecular nitrogen, and minor species below the per cent level: water vapour, carbon monoxide, oxygen and its photodissociation product ozone. We have seen also that water was present in the early history of the two planets in much greater abundances than today.

This similarity suggests that the primitive atmosphere of the Earth must have been also dominated by $CO_2$, with an abundance of $H_2O$ and a minor contribution of $N_2$. The state in which water was present depended upon the surface temperature of the planet; it was thus a function of its heliocentric distance but also the time in its history. In the early days of the planets' history, the Sun's radiation was only about 70 per cent of its present value. As mentioned above, water on Venus was probably in liquid form; in contrast, it should have been in the form of ice on Earth. How did the Earth escape the permanent snowball state? Probably through the release of internal energy from volcanic activity which fed the atmosphere with carbon dioxide and generated enough greenhouse effect to partially melt the ice. Once water happened to be in liquid form on Earth, a self-regulation mechanism took place which transformed carbon dioxide into calcium carbonate under the oceans. Were the temperature to increase slightly, the $CaCO_3$ precipitation would increase too, leading to a slight decrease in the gaseous $CO_2$, limiting the greenhouse effect and stabilizing the temperature. As a result, in contrast with Venus where a runaway greenhouse effect took place, the temperature at the Earth's surface remained more or less constant over its history and the planet retained its water ocean, allowing the development of life in all the forms we know.

The situation of Mars was different in two aspects: being further from the Sun, the planet was colder; and as it is a tenth as massive as the Earth, its internal energy was accordingly reduced, leading to a much weaker atmospheric outgassing, and its lower gravitational field was also unable to capture an atmosphere as dense as that of Venus and the Earth. These two reasons explain why the Martian primordial atmosphere, although probably denser than today, is more tenuous than those of its neighbours. As mentioned above, we have now good evidence that the early Martian atmosphere was warmer, denser and wetter than today. In the first billion years, when the dynamo (generated by internal energy coming from radioactive decay) was still active, volcanic activity must have been sufficient to generate atmospheric outgassing, increasing the atmospheric temperature and allowing abundant liquid water to be present at the surface. Because of the limited amount of internal energy (due to the relatively low mass of the planet), the dynamo somehow ceased, the atmosphere escaped, liquid water disappeared (except probably during transient catastrophic phenomena) and the remaining water was trapped under the surface, under the polar regions and in the permafrost.

## 3.5   WATER ON EARTH: WHERE DID IT COME FROM?

How did terrestrial planets acquire their atmospheres? We have seen that the planets were formed from solid particles of relatively high density, dominated by silicates, oxides and metals. The hydrogen molecule was light enough to escape the planet's gravity field. But some other molecules ($CO_2$, $CO$, $H_2O$ and $N_2$) could be partly outgassed from the interior. It is generally believed, however, that a significant part of the terrestrial planets' atmospheres came from meteoritic, micrometeoritic or cometary impacts, after the planets were formed. This is especially true for water on Earth, mostly found in liquid form in the oceans (Figure 3.17).

An important diagnostic of the origin of the terrestrial oceans is provided by measurement of the D/H ratio in water. We have seen

FIGURE 3.17 Oceans on Earth: the origin of water on our planet is a pending question. For colour version, see plates section. (Image courtesy of NASA.)

above (see Subsection 1.2.3) that its value in the oceans (the VSMOW value) is $1.6 \times 10^{-4}$, which we can compare to other values of the D/H ratio elsewhere in the Solar System and even in the Universe. Indeed, the deuterium abundance is a diagnostic of the formation time and temperature of the body in which it is measured (see Section 1.2). In the solid phase, at low temperatures, the D/H ratio is different from that in the gas phase. Measurements made in the interstellar medium, as well as low-temperature laboratory measurements, show that the D/H ratio increases as the temperature decreases. The lower the temperature, the higher the D/H ratio; this is illustrated, in particular, in comets, for which the formation temperature is only a few tens of kelvins (30–50 K), where the D/H ratio (measured from $HDO/H_2O$) reaches a value of $3 \times 10^{-4}$, i.e. 15 times the protosolar value and two times the terrestrial value. In the case of Uranus and Neptune, which are mostly made of ices ($H_2O$, $CH_4$, $NH_3$, $H_2S$ and others), a D/H enrichment (measured both from $HD/H_2$ and $CH_3D/CH_4$ ratios) is also noticed with respect to Jupiter and Saturn (D/H = $5 \times 10^{-5}$ on Uranus and Neptune and $2 \times 10^{-5}$ on Jupiter and Saturn).

As pointed out before (Section 1.2), these comparisons tell us that a measurement of D/H in the Solar System is an important diagnostic of the formation temperature of the object where it is observed. The D/H value in terrestrial oceans ($1.6 \times 10^{-4}$) indicates that water on Earth cannot have arisen entirely from outgassing: in that case, the D/H ratio would be much lower, close to the protosolar value or even higher, as the formation temperature of the Earth was too high to allow ices to be present. Water in the oceans was most likely brought in by external impactors. Where did they come from? Some scientists suggested that they could have a cometary origin; the implications for astrobiology could be important, as comets might also have brought in organic and prebiotic molecules, and could thus be responsible for the appearance of life on Earth. However, the first measurement of D/H in a comet (it was comet Halley, observed *in situ* at the time of its 1986 passage) led to the value of $3 \times 10^{-4}$ mentioned above, twice the VSMOW value. Other remote sensing measurements performed 10 years later on two other comets, Hale–Bopp and Hyakutake, confirmed this result. In parallel, measurements of D/H in objects from the outer asteroid main belt, obtained from meteoritic measurements, led to the conclusion that this class of objects could be mostly responsible for the terrestrial water. But the story is not over yet: in 2011, new measurements from the Herschel satellite on another comet, Hartley 2, led to a value compatible with the VSMOW value. Interestingly, the origin of Hartley 2 is different from the three previous comets, Halley, Hale–Bopp and Hyakutake, which all come from the Oort cloud in the confines of the Solar System, at about 40 000 AU from the Sun. In contrast, Hartley 2 is a 'Jupiter family comet' which originates from the Kuiper belt, just beyond the orbit of Neptune at 40–100 AU from the Sun. Could this reservoir be the origin of the terrestrial water? Further measurements on other Kuiper belt comets will be necessary to better answer the question of the origin of water on Earth.

## 3.6 EARTH'S COMPANION, THE MOON

A special mention must be made of the Moon, our satellite. Although of little interest for exobiology, the Moon exhibits some unusual features. First, its size relative to the Earth makes it unusual compared with other satellites: the Earth and Moon can be considered as an Earth–Moon system (Figure 3.18), like the Pluto–Charon system in the Kuiper belt. This unusual configuration can be explained in the light of its formation scenario. According to current models, shortly after its formation, the proto-Earth was struck tangentially by a Mars-sized body. Material blasted into Earth orbit reassembled to form the Moon; this theory allows us to explain the angular momentum of the Earth–Moon system and the relatively low density of the Moon ($3.3$ g cm$^{-3}$) compared with the Earth. Dynamical simulations show that the presence of a large satellite around the Earth allowed the system to stabilize and probably contributed significantly to the sustainability of life on our planet.

As a physical body, the Moon is a dead world, heavily cratered, very similar in appearance to Mercury. Its surface is characterized with

FIGURE 3.18 Earth and Moon system as observed by the Galileo spacecraft during its December 1992 flyby. For colour version, see plates section. (Image courtesy of NASA.)

large impact basins, of volcanic origin and filled with basaltic lava, called 'mare'.

Liquid or gaseous water cannot persist at the surface of the Moon: it is promptly dissociated by the solar UV radiation field. Still, over the past years, the discovery of traces of water by several space missions (Clementine, Cassini, Chandrayaan-1) has raised a debate in the community. The water content is, in any case, extremely low, and of little interest for future exploration prospects. It could come from melt inclusions in the upper mantle or originate from reactions of the surface with the solar wind.

Our main interest in the Moon, within the framework of this book, is related to the history of manned space exploration, which started with the first man on the Moon, landed there by Apollo 11 in 1969. Over 50 years since the last Apollo 17 mission in 1972, no further manned exploration of the Moon has taken place. The Moon is still considered as the first step which was necessary for human space exploration (see Chapter 6).

## 3.7 BETWEEN TERRESTRIAL AND GIANT PLANETS, THE ASTEROIDS

We have seen (Chapter 1) that, in the protosolar disk, the process of planetary formation – core accretion – led to the formation of two classes of planets: the terrestrial ones, smaller and denser, closer to the Sun; and, at greater distance, the giant planets, larger, less dense, with their ring and satellite systems. Between the orbits of Mars at 1.5 AU and Jupiter at 5.2 AU, there is no planet. The currently accepted explanation is that in this region, the formation of a planet as big as Mars was inhibited by the gravitational perturbations induced by Jupiter. However, there was matter in the protoplanetary disk, gas and dust; the gas dissipated at the time of the T-Tauri phase, but the solid particles accreted in small bodies which were fragmented by mutual collisions: they are the asteroids of the main asteroid belt. The biggest ones, Ceres, Pallas, Juno and Vesta, were discovered at the beginning of the nineteenth century; their diameters range between 500 and 1000 km and are thus comparable in size with many outer satellites. The smallest observed asteroids (those coming close to the Earth, or those encountered by spacecraft) have sizes less than a kilometre (Figure 3.19).

With a population of about 450 000 numbered objects, the main belt is by far the major reservoir for asteroids, but other families also exist: closer to the Earth, the Apollos, the Amors and the Atens with about 6000 objects identified; and on the two so-called 'Lagrangian points' L4 and L5 of Jupiter's orbit, with about 4000 objects known. Beyond the orbit of Neptune, the trans-Neptunian objects (TNOs) are a new class of small bodies. Although their existence had been predicted decades ago from dynamical simulations, the first identification was made by two astronomers from the University of Hawaii, Dave Jewitt and Jane Luu in 1992. It was then recognized that Pluto was the first member (and among the largest) of this new family. About 1300 TNOs

FIGURE 3.19  The asteroid Ida and its satellite Dactyl, as imaged by the Galileo spacecraft in its way to Jupiter on 18 August 1993. This was the first direct observation of an asteroid with its own satellite. (Image courtesy of Galileo Project/JPL/NASA.)

have been identified today. Between the TNOs and asteroids, there is another class of intermediate objects, the Centaurs, with mean heliocentric distances ranging between the orbits of Jupiter and Neptune; these might be TNOs that had their orbits perturbed by gravitational perturbations by Neptune and possibly the other giant planets.

The main difference between asteroids and comets is the lack of gaseous activity in the former case. This is certainly true for main belt and near-Earth objects. In the case of Centaurs, however, the frontier is less clear. Centaurs, indeed, may show cometary-type activity when they come close enough to the Sun. We know about a hundred objects of this kind; among them are 2060 Chiron, the first identified Centaur (also named 95 P/Chiron to account for its double nature), 5145 Pholus and 10199 Charilko, the biggest one known, with a diameter of 260 km. Saturn's satellite Phoebe might be a captured TNO. Clearly, a strong link exists between the different categories of small bodies in the outer Solar System: Centaurs, comets, outer satellites and TNOs.

Our knowledge of the physical and chemical properties of asteroids mostly relies on remote-sensing imaging and spectroscopy, from the visible to the far-infrared range. In addition, a few asteroids have been encountered by spacecraft, in particular Gaspra (1991) and Ida (1993) by Galileo, Mathilde (1997) and Eros (2000) by NEAR, Braille (1999) by Deep Space 1, Masursky (2000) by Cassini, and Itokawa (2005) by Hayabusa. Monitoring the light-curves gives information about the shape, the rotation period, and the inclination of the spin axis; near-infrared spectroscopy gives access to the mineralogy of the surface. In the case of unresolved objects, simultaneous observations in the visible and mid- or far-infrared range allow the diameter and the albedo to be obtained. The albedos of asteroids are widely distributed, from a few per cent in the case of carbonaceous objects up to 10–30 per cent for silicated objects. Densities, in contrast, are more difficult to determine. Measurements have been obtained on about 40 bodies, either from a spacecraft or from occultation measurements in the case of double systems. The mean density appears to be close to $1 \, \text{g} \, \text{cm}^{-3}$, the density of water, which strongly suggests a porous structure. A typical example is Itokawa (Figure 3.20), imaged by the

FIGURE 3.20 The asteroid Itokawa as viewed by the Japanese spacecraft Hayabusa in September 2005. In spite of some technical problems, the spacecraft collected samples which were returned to Earth in June 2010. Itokawa is an S-type, Apollo and Mars crosser asteroid. Its unusual shape and surface structure suggests that it might be a 'rubble pile' formed from the accretion of two or more fragments. (Image credit: ISAS/JAXA.)

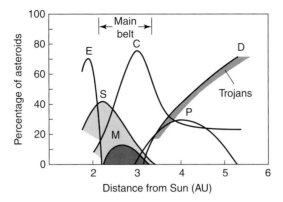

FIGURE 3.21 The distribution of different types of asteroids as a function of the heliocentric distance. The E, R and S types dominate in the vicinity of the Sun, where the formation temperature was higher. The more primitive types (C, P and D) dominate at larger heliocentric distances, beyond the main asteroid belt. (From NASA's Cosmos, http://ase.tufts. edu/cosmos/ © 2010, Professor Kenneth R. Lang, Tufts University.)

Hayabusa spacecraft, which exhibits a very irregular surface apparently composed of blocks of different sizes.

Asteroids have been divided into different taxonomic classes on the basis of their mineralogy, derived from their visible and near-infrared spectra (Figure 3.21). The dominant class is the S-type (about 40 per cent) rich in metals and silicates. The next group (about 30 per cent) is the C-type, rich in carbon, organics and hydrated silicates. Among the other minor classes, the M-type (2 per cent) is dominated by metals and the D-type (about 2 per cent) are made of carbon and organic-rich silicates. Asteroids of interest for astrobiology are obviously the C, B, D and P types where carbon and organics are present, which might have fed the Earth with organic and possibly prebiotic molecules. In addition, there is some evidence that hydrated minerals are also present in many M-type asteroids, so these objects might also have contributed to feed the Earth with water.

It is interesting to observe the distribution of asteroidal types as a function of their heliocentric distance. Not surprisingly, the S and M types dominate in the main belt and the inner Solar System, whereas the C, P and D types dominate outside the main belt. In particular, the D-type abundance increases at 4 AU and beyond.

Could the water of the Earth have been brought in by asteroids? The debate is still open. Two models have been proposed: either water was incorporated at the time of the Earth's formation, or it arrived at a later stage as a 'veneer' brought in by cometary and asteroid impacts. As discussed above, the D/H ratio is a key diagnostic to test these two models. The existing measurements of cometary D/H seem to exclude Oort comets as possible parents for terrestrial water, but the answer is not clear if we consider Kuiper belt comets, and more measurements are needed. On the other hand, the D/H ratio measured in meteorites whose parent bodies belong to the outer main belt appears in agreement with the VSMOW value. In addition, compositional measurements of these samples show that they contain up to 10 per cent of water by mass. It is thus possible that planetary embryos coming from the outer main belt were incorporated into the Earth at the time of its formation and contributed to the oceans (see Section 3.5).

In the case of organics, however, their incorporation at the time of the Earth's formation seems unlikely, as complex organics should not have survived on Earth until the planet cooled down. One could thus imagine that organic molecules came to Earth as a result of meteoritic bombardment from both comets and carbonaceous asteroids after the planet's formation phase, with a possible peak at the time of the Late Heavy Bombardment. This leads us to the realm of the outer Solar System and its possible interaction with the terrestrial planets.

# 4    Searching for habitable sites in the outer Solar System

Looking for habitable conditions in the outer Solar System leads us to the natural satellites rather than the planets themselves. Although the theoretical conditions under which life might be sustained on natural satellites are similar to those of planets, there are key environmental differences which can make moons of particular interest in the search for extraterrestrial life. The gaseous giant planets cannot provide even the minimal conditions of a surface or interior with suitable pressures and temperatures to sustain life. But the moons around these planets offer a great range of possibilities for exploring habitability conditions and furthermore studying the question of the emergence and evolution of habitable worlds in our Solar System, in some cases more so than any other object closer to the Sun. Scientists generally consider the probability of life on natural satellites within the Solar System to be remote, though the possibility has not been ruled out.

Within the Solar System's traditional habitable zone, the only candidate satellites are the Moon, Phobos and Deimos, and none of these has an atmosphere or water in liquid form. But, as discussed in Chapter 2, the habitable zone may be larger than originally conceived. The strong gravitational pull caused by the giant planets may produce enough energy to sufficiently heat the cores of orbiting icy moons. This could mean that some of the strongest candidates for harbouring extraterrestrial life are located outside the solar habitable zone, on satellites of Jupiter and Saturn. The outer Solar System satellites then provide a conceptual basis within which new theories for understanding habitability can be constructed. Measurements from the ground and also from the Voyager, Galileo and Cassini spacecraft have revealed the potential of these satellites in this context, and our understanding of habitability in the Solar System and beyond can be greatly enhanced by investigating several of these bodies together.

Objects with oceans in our Solar System may belong to one of two families: 'terrestrial' planets, or icy satellites of giant planets. The question is whether the liquid water therein has existed for periods of time long enough to have been biologically useful, because a frozen surface may not necessarily have inhibited life. If liquid layers exist below ice layers and these water-reservoirs are in contact with heat sources from the interior of the planet (from radioactive decay, volcanic interactions or hydrothermal activity) the planet may be considered as a potential habitat.

In particular, the satellites around Jupiter and Saturn, closer to the Sun than those of the ice giants, have been revealed as extremely interesting active bodies from the viewpoint of astrobiology. Indeed, several of them show promising conditions for habitability and the development and/or maintenance of life. Europa, the smallest of Jupiter's four Galilean satellites, may be hiding, under its icy crust, a putative undersurface liquid water ocean which may be in direct contact with a silicate mantle floor and kept warm through time by tidally generated heat. If its mantle is geologically active like that on Earth, giving rise to the equivalent of hydrothermal systems, the simultaneous presence of water, geodynamic interactions, chemical energy sources and a diversity of key chemical elements may fulfil the basic conditions for habitability.

Further out, around Saturn, Titan constitutes another such example. The largest Kronian moon (Kronos being the Greek name for Saturn), Titan is the only other body in the Solar System besides Earth to possess a dense atmosphere composed essentially of nitrogen (~98 per cent), which combines with methane (~1.5 per cent in the stratosphere) to give rise to a host of organic compounds. The presence of seasonal effects manifesting themselves in the atmosphere, geomorphological features similar to our planet's, and a more and more probable internal liquid water ocean make Titan one of the most astrobiologically interesting bodies, as revealed by the ongoing Cassini–Huygens mission.

Future extensive and *in situ* exploration concepts could help us improve our understanding of Titan and Europa, and some of their siblings such as Ganymede or Enceladus. For all these reasons, these moons have often been proposed as the main targets for future missions.

In the outskirts of our Solar System we find Uranus and Neptune surrounded by moons which are also unique in many aspects, such as Triton and its geysers. Hereafter we investigate the astrobiological prospects of this extended habitable zone, focusing on the smaller bodies which are being more and more revealed as exciting objects in the search for habitable conditions.

Although some 50 or so extrasolar moons have recently been announced, there is still no firm confirmation or perception of how common they may be, what their characteristics are and how many could be considered habitable. Some studies do, however, consider exomoons as among the most likely habitable environments, as we will see in Chapter 5.

Furthermore, natural satellites have been considered by some astrobiologists as potential candidates for space colonization if the right artificial environment could be created. As Coustenis and Taylor argue, 'the most Earth-like moon in the Solar System is Titan', with a nitrogen atmosphere, organic chemistry and extents of liquid hydrocarbons on its surface, but its surface nevertheless remains a hostile place for life (Coustenis and Taylor, 2008). Terraforming of moons might be considered but is outside the limits of current technology, as we shall discuss in Chapter 6.

Further out in the Solar System we find more objects of interest, such as the comets and the trans-Neptunian objects. Cometary atmospheres incorporate prebiotic molecules and pristine material that help us to understand how the Earth's atmosphere came to be, and also how the Sun's neighbourhood formed and evolved. Furthermore, comets are filled with water...

Triton (orbiting around Neptune) and objects beyond Neptune's orbit have been hypothesized to harbour subsurface liquid water oceans (even today, and even at the very low temperatures of their environments). Thus, following the water is not a wild goose chase in the outermost regions of the Solar System and even beyond, as we shall see in the next chapter.

## 4.1    THE OUTER SOLAR SYSTEM: A HUGE RESERVOIR OF FROZEN WATER

Even if they are located beyond the so-called snow line (or ice line; the distance from the Sun beyond which water cannot be liquid on the surface), large satellites of gas giants are known to include substantial quantities of water. Indeed, with an average density around 1.8 g cm$^{-3}$, the largest moon is composed of almost 45 per cent by weight of water. Among these satellites, some have a silicate-rich core and may contain underground liquid water oceans deposits in contact with it (as for Europa) and therefore harbour heat and chemical sources below their ice crust. This is important as on Earth in such a context we find life (in hydrothermal vents at the bottom of oceans). Other satellites may have liquid water layers encapsulated between two ice layers, or liquids above ice. In the study of the emergence of life elements on such satellites, the timescale is of the essence. If this liquid environment persists for long enough, it might see the emergence of life; but, on the other hand, this could be inhibited if the ocean is so isolated from the surrounding environment that it is impossible to assemble the concentration of ingredients necessary for life or the proper chemical inventory for the relevant biochemical reactions.

Icy layers on the outer moons can be very thick (Figure 4.1). One can assume their thickness to be of the order of a few hundred km for Ganymede, Callisto and Titan. As Galileo data indicate, unlike Ganymede, Callisto is probably not fully differentiated, which means that the icy layer might be thicker, but mixed with silicates. For Europa, a thinner icy layer (possibly 100 km) is hypothesized, because its proportion of water is 'only' 10 weight per cent. For Titan, an icy crust some hundred kilometres in depth is expected. Understanding the internal structure of the water layer requires knowledge of the water phase diagram under certain pressure and temperature conditions.

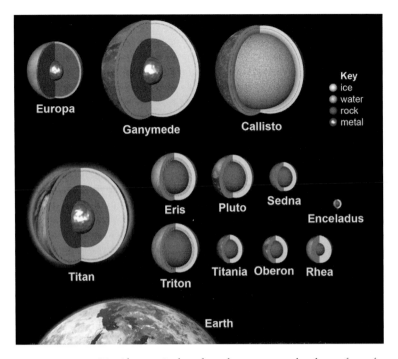

FIGURE 4.1 Liquid water is thought to be present under the surface of many bodies in our Solar System, particularly the Galilean moons of Jupiter, such as Europa, Callisto and Ganymede, and Saturn's Titan and Enceladus. In addition, models of heat retention and heating via radioactive decay in smaller icy bodies suggest that Saturn's Rhea, Uranus' Titania and Oberon, Neptune's Triton, trans-Neptunian objects and dwarf planets Pluto, Eris, Sedna and Orcus may also have oceans underneath solid icy crusts approximately 100 km thick. The models vary and in some cases predict that the liquid layers are in direct contact with the rocky core, which allows efficient mixing of minerals and salts into the water, while in others layers of high-pressure phases of ice are thought to underlie the liquid water layer. For colour version, see plates section. (Image credit: Doug Ellison, for the Planetary Society.)

On Earth, almost all of the ice present in the biosphere is in the form of a hexagonal crystal constituting what we are used to calling frozen water, but whose scientific term is 'ice Ih' (pronounced 'ice one aitch'). Ice Ih is stable down to –200 °C (73 K; –328 °F) and can exist at pressures up to 0.2 GPa. It exhibits many peculiar properties which are relevant to the existence of life and regulation of global climate. If the temperature is between 250 and 273 K, it is possible for a liquid layer to be located below an ice Ih layer, because ice Ih is less dense than the liquid. This situation is compatible with large icy moons and indicates that many ice polymorphs can exist in the pressure range 0–2 GPa, relevant to icy satellites. As noted in recent thermodynamical studies, the icy moons of Jupiter and Saturn can benefit from the fact that the melting curve of the low-pressure ice Ih decreases with pressure. Like the recently discovered exoplanets, they could, according to some studies, in the case of a coupled sea/ice system provide the necessary conditions for life emergence as on the primitive Earth.

As explained in Subsection 2.3.2 and following the suggested classification of habitable worlds proposed in Lammer *et al.* (2009), Class I habitats are similar to the Earth and generally represent bodies whose stellar and geophysical conditions allow them to evolve so that complex multicellular lifeforms may emerge. Class II habitats are planets or satellites on which life may indeed emerge and develop, but owing to different stellar and geophysical conditions they evolve toward Venus- or Mars-type hostile environments where complex lifeforms may not develop. Class III habitats are planetary bodies where subsurface water oceans exist, like on Europa, but where they also interact directly with a silicate-rich core, while class IV habitats have liquid water layers sandwiched between two ice layers, or liquid extents above ice (Figure 4.2).

In this chapter we concern ourselves more with the two latter types of habitats. In the Lammer *et al.* (2009) classification, 'Class III habitats where subsurface oceans are in contact with silicates on the sea-floor open the question of where the building blocks for life could come from.' If the organic material necessary to start life is supplied by exogenous sources such as meteoritic and cometary impacts and by precipitation of interplanetary dust, it still has to find its way into the subsurface oceans. As Lammer and colleagues point out, in addition:

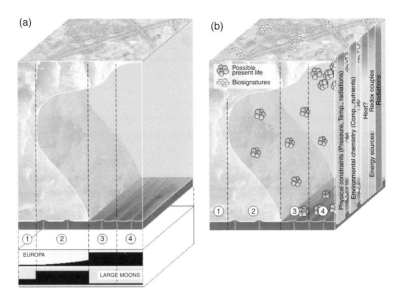

FIGURE 4.2 (a) Possible locations of liquid layers in the icy moons of Jupiter are plotted here as a function of depth: (1) completely frozen; (2) three-layered structures impeding any contact between the liquid layer and the silicate floor; (3) thick upper icy layer (>10 km) and a thick ocean; (4) very thin upper icy layer (3–4 km). (The four different sections apply to different structures for the moons, and are joined here only for the purposes of the illustration.) The black shading at the base indicates probability of finding the structure above. Structures 3–4 are the most probable for Europa. The larger moons Ganymede and Callisto are located in the left region (1 or 2) where internal pressures are sufficient to allow for the formation of high-pressure ice phases. Oceans in Ganymede and Callisto – if they exist – should be enclosed between thick ice layers. For colour version, see plates section. (Redrawn from Lammer *et al.*, 2009.) (b) Current habitability of Europa. Possible locations of present life and biosignatures have been plotted as a function of depth. Habitability depends on physical and chemical constraints which are indicated on the right using colour scales (green: highly favourable; red: hostile). Numbers refer to possible interior structures described in (a). For colour version, see plates section. (Redrawn from Blanc *et al.*, 2009.)

this material has to reach meaningful concentrations in some (small) compartment of the ocean, which is hard to imagine in a connected body of water as large as a planet-wide subsurface ocean. However, one should keep in mind that synthesis of organic material by either Fischer–Tropsch reactions (which convert carbon monoxide and hydrogen into hydrocarbons) or catalytic cycles are possible under the high pressure/high temperature conditions occurring at deep-sea vents.

The terrestrial hydrothermal vents are environments highly favourable for life, contrary to what might be expected in the absence of sunlight. When in contact with silicates, the hot water from the hydrothermal vent can extract the minerals and hence the C,H,N,O,P,S (CHNOPS) that are necessary and beneficial to the large community of living organisms around them. These species feed on organic material produced by a form of bacteria that use energy obtained by oxidizing reactive chemicals (such as hydrogen). This process, called chemosynthesis, is a substitute for photosynthesis and leads to the assumption that life can indeed exist in subsurface oceans, provided liquid water and some sort of energy are available. Anaerobic chemosynthetic bacteria might then exist on outer Solar System moons. As a consequence, one could imagine that life might exist, as on Earth, in the vicinity of such environments (hydrothermal vent-like formations in subsurface oceans in contact with silicates) in the outer Solar System satellites, if reduced radicals are found in the mineralized water and the environment can also provide the energy to sustain the living organisms, although probably not to power them so they can grow, develop and evolve. This process is more easily envisioned on Enceladus and Europa than on Ganymede and Callisto.

Indeed, the latter two Jovian satellites are considered as Class IV habitats, and such habitats can be found not only in the Solar System but in exoplanets as well. These are environments where a water ocean may be contact with a thick ice layer. In that case, as Lammer and co-authors point out, 'a much better situation for the influx of organic material from outside the planet' is afforded for such bodies. The main problem for Class IV habitats, however, is to assemble the necessary ingredients for life in a given location, difficult for a planet whose surface is covered in a deep water ocean with nothing to act as a 'magnetic centre' for organic chemistry, in particular at the right temperature conditions. To better understand this problem and to test the possibilities, we can envisage laboratory experiments and *in situ* studies in cold regions of the Earth such as the deep isolated lakes in Antarctica, as the physical properties there resemble those found on some of these moons.

But all in all, Class III and Class IV waterworlds appear to have considerable astrobiological potential and could be possible habitats if some conditions are combined. Besides the correct pressure and temperature conditions for the presence of liquid water, the planetary object should have adequate energy sources to support metabolic reactions, and chemical elements to provide nutrients for the synthesis of biomolecules. So another very important aspect to consider when looking at the habitability of outer moons is the tidal effect, which can be a significant source of energy on natural satellites and an alternative energy source for sustaining life. Moons orbiting gas giants or brown dwarfs are likely to be tidally locked to their primaries: that is, their day is as long as their orbit. Although tidal locking is expected to affect planets within habitable zones adversely by producing an inhomogeneous distribution of stellar radiation, it may also have favourable effects on satellites by exposing them to powerful tidal forces, inducing plate tectonics through the stresses imposed on the lithosphere and causing higher temperatures. This could bring on volcanic activity, which regulates a moon's temperature, and could even create a geodynamo effect that would provide it with a significant magnetic field.

Monoj Joshi, Robert Haberle and their colleagues modelled the temperature on tide-locked exoplanets in the habitability zone of red dwarfs. They found that an atmosphere with a carbon dioxide pressure of only 1 to 1.5 atmospheres not only allows habitable temperatures but also allows for liquid water to exist on the dark side. If a moon is tidally locked to a gas giant it suffers large temperature variations, but these are less extreme than those for a planet locked to a star. There are ways to bypass the temperature issue, since even small amounts of carbon dioxide could make the thermal properties of a planet suitable for life, as we shall see in Chapter 5.

As a consequence, some of the Saturnian and Jovian satellites, such as Ganymede, Europa, Titan or Enceladus, have been identified as potentially habitable environments and targets for our space-based and ground-based search for life (or at least for habitable conditions). In the Solar System's neighbourhood, such potential habitats can only be properly investigated with appropriately designed space missions; for extrasolar planets we cannot do that, so we must rely on remote sensing, and thus living organisms whose existence does not affect or have signatures in the atmosphere of their host planet will not be detectable. Hereafter we look more closely at these bodies.

## 4.2   JUPITER'S SATELLITES

The satellites of the Jovian system are unique bodies, each different and precious in its own way. In the case of the Jupiter system, three of the Galilean moons, Ganymede, Europa and Io, are locked in a 1:2:4 orbital Laplace resonance. The Laplace resonance (a resonance is explained in Subsection 1.2.2) involves three or more orbiting bodies; other Laplace resonances exist, as in the Gliese 876 system. The resonance is important because it means that these objects periodically exert a strong gravitational influence on each other.

Ganymede is the biggest moon of the Solar System, even bigger than the planet Mercury. Gravity data point to a fully differentiated structure. It possesses a very tenuous atmosphere, a hydrosphere which may be at least 500 km thick (meaning 50 weight per cent), a silicate mantle and an iron core. A liquid layer up to 100 km thick is trapped between the icy crust on top and a layer composed of high-pressure ices, which do not exist in natural environments on the Earth. It is one of three solid bodies that possess an intrinsic or dynamo magnetic field (the others are Mercury and the Earth), and this Ganymede magnetic field is in turn embedded in the Jovian magnetic field. Finally, it also has an induced magnetic field, generated by currents flowing in a conducting liquid, the best proof of the existence of the deep ocean.

Europa is layered in a similar way. We think that it possesses a similar metallic core and a silicate layer, about the same size as Ganymede's, but a thinner hydrosphere no more than 100 km thick. The big difference here is that the liquid layer is almost certainly in direct contact with the silicate ocean floor.

Callisto is very similar to Ganymede, apart from the fact that it is not fully differentiated and that it does not have an intrinsic field. This lack of differentiation, rather surprising for a body of this size, may be due to the fact that Callisto is decoupled from the Laplace resonance (as defined above), one of the most striking pieces of evidence of the complexity of the coupling processes occurring in the Jovian system. Its surface is very old and heavily cratered, and finally, it probably also possesses a deep ocean similar to Ganymede.

These icy Galilean satellites, Callisto, Ganymede and Europa, thus present a great diversity of surface features probably caused by their different evolution, despite the fact that they are located in the same 'neighbourhood'. The unique features of their geology are a testimony to their formation parameters, such as composition, density and temperature, as well as evolutionary factors such as geophysical processes and the stage of differentiation. In addition, space weather – for instance interactions with the Jovian magnetosphere and stellar wind – and tidal effects have left their marks on the landscapes we see today (Figure 4.3). Thus, by observing the surface features, one finds evidence for the actual or past presence of aeolian (wind-driven) processes, cryovolcanism, impact cratering and other activity. The Galilean satellites show signs of an increase in geological activity with decreasing distance to Jupiter. Io, nearest to Jupiter, is obviously volcanically active. Europa could still be tectonically and volcanically active today, whereas Callisto, the outermost Galilean satellite, may be geologically 'dead'. Understanding the gravitational interactions between Jupiter and the Galilean satellites is essential for many aspects of Jupiter system science, including habitability, through their influence on the evolution of a satellite's interior and surface. In particular, the evolution of the Laplace resonance may be important

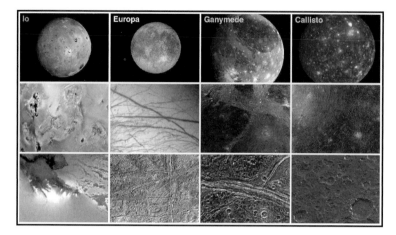

FIGURE 4.3 The surfaces of the Galilean satellites. The image for Europa shows a region of its crust made up of blocks which are thought to have broken apart and 'rafted' into new positions. These features are the best geological evidence so far that Europa may have had a subsurface ocean at some time in its past. Combined with the geological data, the presence of a magnetic field leads scientists to believe an ocean is most likely present at Europa today. In this false colour image, reddish-brown areas represent non-ice material resulting from geological activity. White areas are rays of material ejected during the formation of the 25-km diameter impact crater Pwyll (see global view). Icy plains are shown in blue tones to distinguish possibly coarse-grained ice (dark blue) from fine-grained ice (light blue). Long, dark lines are ridges and fractures in the crust, some of which are more than 3000 km long. These images were obtained by NASA's Galileo spacecraft during September 1996, December 1996 and February 1997 at a distance of 677 000 kilometres. For colour version, see plates section. (Image courtesy of NASA/JPL/DLR.)

for the subsurface oceans of Europa and Ganymede, and for the future of volcanism on Io, on which tidal dissipation can be an important heat source, as it can for some of the other satellites.

Voyager and Galileo data indicate that Europa and Ganymede, and possibly Callisto, possess important prerequisites to be considered habitable. Galileo's detection of induced magnetic fields, combined with imaged surface characteristics and thermal modelling of the moons' evolution, advocates the presence of liquid water oceans below the icy crusts of Ganymede, Europa and Callisto. These three satellites are the targets of the first large mission being developed by ESA in the framework of the Cosmic Vision 2015–2025 framework, and their properties as known today and as will be studied in the future are described in the report of the European JUICE (Jupiter Icy Moons Explorer) space mission (see Subsection 4.2.3) (JUICE, 2011, p. 29). According to that report:

> the depth and composition of the oceans, as well as the dynamics and exchange processes between the oceans and the deep interiors or the upper ice shells, remain unclear. Furthermore, it is unknown whether liquid water reservoirs or compositional boundaries exist in the shallow subsurface ice and how the dynamics of the outermost ice shell is related to geologic features and surface composition.

There is ample evidence that ice and liquid layers both exist on these moons, but there is no certainty as to their exact composition. Observations indicate that the water in the subsurface is most probably mixed with other components such as the salt hydrates or carbon dioxide that we find on the surfaces of most of the Galilean satellites. Besides the basic biochemical components that are required for life (CHONPS; see Chapter 2), other species, such as sodium, magnesium, calcium, iron and potassium, also play an important role. However, if these constituents are found on a surface, the conditions for their survival on Europa will not be the same as on Ganymede, as the distance to the planet and the received radiation doses are very different. Close to Jupiter, as in the case of Europa, this radiation is extremely harmful and probably fatal to any organics or any other biota present there. At the very least, the surface organics and minerals on Europa's surface would suffer from large alteration processes. This is a pity, because radiation also helps to produce oxidants on the surface (molecules with elements in a high oxidation state) through redox reactions (see Chapters 1 and 2), which are essential in life, although harmful to our bodies in high quantities. The composition of the surface is largely reflected in the tenuous atmospheres (and exospheres) of the Galilean satellites observed essentially in UV measurements from space and from the ground, which are the result of sputtering and sublimation of the surface materials. Conversely, the atmospheric constituents also influence the surface composition. Europa has a dioxygen atmosphere with traces of sodium and potassium. Ganymede also has a thin oxygen atmosphere and a hydrogen exosphere.

The oxidants present on the surface could, in some cases, if the ice layer is permeable or not too thick, penetrate down to the ocean, thus enhancing the habitability potential. Similarly, if an undersurface ocean exists under a significant ice layer (thus protected from the radiation effect), the habitability potential is preserved.

Besides their interiors, the Galilean satellites are interesting objects because of their diverse surface conditions (Figure 4.3). In the case of the large moons Ganymede and Callisto, observations show old, densely cratered terrains with a wide spectrum of sizes and multi-ring structures. The surface morphology on Europa is very different, with few craters, suggesting a relatively young surface. In addition, the surface composition may be reflecting internal and external processes. The presence of an intrinsic magnetic field around Ganymede is valuable for habitability, as (rather like on Earth) it protects the satellite from highly energetic particles near the equator.

The dimensions, composition and habitable potential of the putative subsurface liquid water oceans on any of these satellites depend enormously on their geological activity, global composition and interior layering. The latter is still mostly unknown, as is the global distribution of the icy versus non-icy material, but it has strong bearings on the origin and evolution of these bodies, as well as on their astrobiology. As mentioned in Chapter 2, the essential difference in classes of habitable worlds lies in the way the water content may or may not be in contact with the silicates in the interior (an important aspect when sources of biological material are considered). Whereas Europa's internal water ocean has a good chance of being in contact with the rocky/silicate layer, those of Ganymede and Callisto are probably not (although in the case of Ganymede it has been argued that silicates could be present).

Hereafter we focus on two of the satellites in the Jovian system that have attracted most of the attention of astrobiologists, namely Europa and Ganymede.

### 4.2.1 Europa

Europa is a Class III habitat unique in the Solar System because of its high rock/ice mass ratio, in that it is the only satellite on which a large ocean might be in contact with the silicate layer. On the other moons, the existence of an ocean implies the occurrence of a very thick high-pressure icy layer at the bottom, which impedes the contact of the liquid with silicates.

The putative existence of a subsurface ocean of course depends on the ability to support liquid water (Figure 4.4). In the case of Europa, this possibility has been suggested based on measurements by the Galileo mission of the induced magnetic field and on the interpretation of geological features. If such an ocean exists, we cannot yet precisely determine its depth or its location. The question of whether there is ice convection within the icy crust, leading to the possibility of an exchange of material between the surface and the ocean, remains open in the thin crust model because of the large number of tectonic features such as cracks and faults visible on the surface (Figure 4.5). This is an important issue for habitability, because Europa's surface is subject to heavy bombardment by energetic particles from Jupiter's radiation belts, leading to the breaking of water molecules and to the generation of hydrogen peroxide, $H_2O_2$, a strong oxidizing component which may be a source of chemical energy. Indeed, $H_2O_2$ tends to decompose exothermically into water and oxygen gas, and it has been detected on Europa's surface in infrared and ultraviolet wavelength spectra.

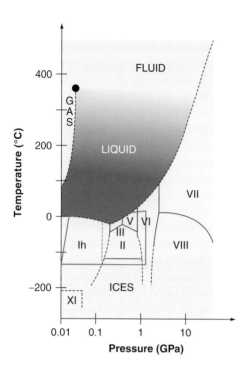

FIGURE 4.4 Phase diagram of the pure water system. Five ice polymorphs can exist in the pressure range relevant to icy moons (box). The dashed line symbolizes the highest pressure that can occur in the water layer of Europa. (From Lammer *et al.*, 2009.)

The determination of the surface topography and the internal structure at the interface between the silicate core and the liquid, together with the detection of any possible mass anomalies there, would allow us to test the hypothesis of whether volcanism and/or hydrothermal activity may exist or have existed, as it does at the Earth's mid-oceanic ridges. Such activity releases into the ocean a variety of organic components that are essential in sustaining possible simple lifeforms on the ocean floor, like those discovered around the Earth's deep-sea hydrothermal vents more than three decades ago.

Given the existence of an ocean, the question of habitability can be summarized by an inspection of the 'triangle of habitability' (Figure 4.6a). In addition to an ocean lasting for a billion years, it involves the presence of the key chemical elements for life (CHNOPS...), and of energy sources. In reviewing the possible

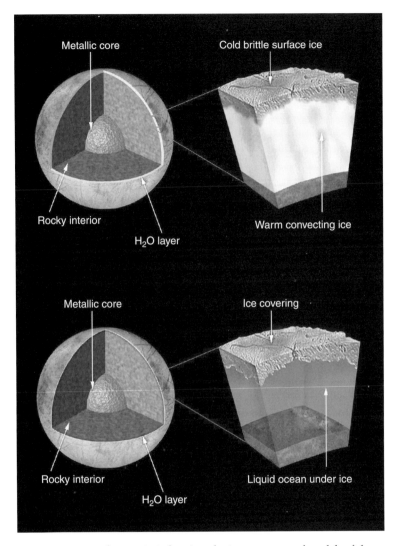

FIGURE 4.5 These artist's drawings depict two proposed models of the subsurface structure of the Jovian moon, Europa. Galileo's observations of surface features can be explained in two ways: either by the existence of a warm, convecting ice layer, located several kilometres below a cold, brittle surface ice crust (top), or by a layer of liquid water with a possible depth of more than 100 km (bottom). For colour version, see plates section. (Image courtesy of NASA/JPL.)

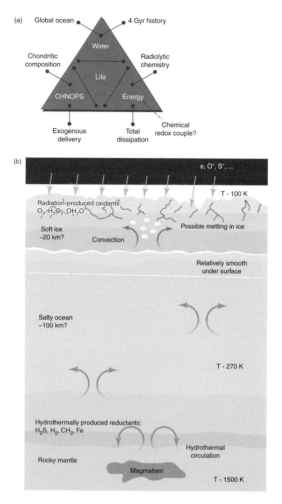

FIGURE 4.6 (a) Illustration of the 'triangle of habitability' for Europa. (Adapted from Hand *et al.*, 2009.) (b) Scheme showing the possible existence of chemical energy sources at Europa. (Adapted from Stevenson, 2000, p. 1823.)

presence and likely abundances of these key elements in Europa's ocean and sea floor, recent studies reached a rather positive conclusion. Concerning the energy sources, the presence of a $H_2O_2$ oxidizing chemistry at the surface, owing to the radiolysis of ice by radiation belt particles, and of a reducing chemistry at the ocean floor if hydrothermal activity exists, opens up the possibility of a redox couple acting at the scale of the ocean (Figure 4.6b). But this only works if oxidants from the surface can feed into the ocean via a (hypothesized) partly permeable icy crust. Further exploration of Europa *in situ* by an orbiter and, better still (when the technology becomes available) by a lander, appears mandatory, as we shall see in Subsection 4.2.3).

## 4.2.2 Ganymede

Ganymede is in many aspects recognized as a unique satellite in the Solar system because of its size, intrinsic magnetic field and geological features. It is believed to be composed of approximately equal amounts of silicate rock and water-ice. Investigations show it to be a fully differentiated body with an iron core (Figure 4.7). Its surface has two

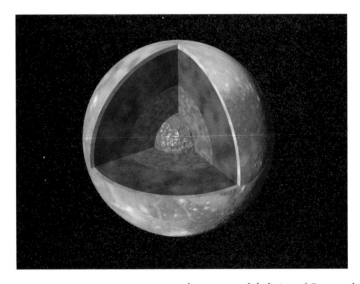

FIGURE 4.7 Voyager images are used to create a global view of Ganymede. The section shown here reveals the interior structure of this icy moon. This structure consists of four layers based on measurements of Ganymede's gravity field and theoretical analyses using Ganymede's known mass, size and density. Ganymede's surface is rich in water-ice, and Voyager and Galileo images show features evidencing geological and tectonic disruption of the surface in the past. As with the Earth, these geological features reflect forces and processes deep within Ganymede's interior which has a differentiated structure with a large lunar-sized 'core' of rock and possibly iron overlain by a deep layer of warm, soft ice capped by a thin, cold, rigid ice crust. (Image courtesy of NASA/JPL.)

main types of terrain. A third of the surface is made up of dark old regions, exposing a high impact crater record and dating back to four billion years ago. The other two-thirds of the geological regions are bright terrains, probably more recent ones indicating resurfacing processes, forming a network and crosscut by extensive grooves and ridges. The cause of the light terrain's special discontinuous geology is not fully understood, but it could be the result of tectonic activity due to tidal heating. As the JUICE Yellow Book report (JUICE, 2011) indicates, we do not currently know if exchange processes are possible between the silicates and the liquid layer in the saltwater ocean

> believed to exist nearly 200 km below Ganymede's surface, sandwiched between layers of ice. [Nevertheless] in the Jovian satellite system Ganymede holds a key position in terms of geologic evolution because it features old, densely-cratered terrain, like most of Callisto, but also widespread tectonically resurfaced regions, similar to most of the surface of Europa.

Unique in some ways but far from rare in others, Ganymede is, indeed, one among many other bodies of its type, both in the Solar System and probably beyond. Ganymede-like objects are expected to be more common in the Galaxy than Earth-like planets since they occupy the giant stellar zone beyond the snow line. Depending on how many such objects we find in the Universe, the probability of having habitable worlds in the Universe (one of the key parameters of the Drake formula for astrobiologists) could be significantly changed.

With the different processes that have marked its atmosphere (which includes O, $O_2$ and possibly ozone) and its surface (such as tectonics, cratering and cryovolcanism), Ganymede is an archetype for understanding many of the icy satellite processes throughout the outer Solar System and beyond, the class of the so-called 'waterworlds', with properties very different from terrestrial planets but still eligible for habitability.

Hydrated minerals have been detected on Ganymede's trailing hemisphere (that is, the side of the moon that faces the direction of the motion as it orbits Jupiter, centred at 270° W, while the leading side is the opposite). In addition, various non-water-ice materials have also been identified on Ganymede's surface from Galileo data and ground-based observations: carbon dioxide, sulfur dioxide, molecular oxygen, ozone and various organic compounds probably formed locally thanks to the radiolytic processes we mentioned above. And since $O_2$ gas is also formed during this process, its detection near the equator is a proof of the existence on the surface of organic material.

As mentioned above, Ganymede is the only satellite in the Solar System known to possess a magnetosphere, thought to be the result of convection within its liquid iron core. This magnetosphere lies in a complicated context: it is embedded in Jupiter's enormous magnetic field and appears as local perturbations of the field lines, interacting with the plasma flow and electromagnetic fields of the Jovian magnetosphere.

### 4.2.3  Future exploration of Jovian satellites

From what has been discussed above, we have established the interest in the exploration of habitats in the satellites of the giant planets. One of the major issues of future missions would be to determine how much water exists in the Jovian system, how it is distributed and what is its content in biological material on and underneath the surfaces of the icy moons, and how the material is transported among the moons by volcanism, sputtering and impacts. As the JUICE report (2011) points out, 'adequate experimentation might also allow us to infer environmental properties such as the pH, salinity, and water activity of the oceans and will investigate the effects of radiation on the detectability of surface organics'.

The JUICE mission was recently selected by ESA as the first large mission within the Cosmic Vision 2015–2025 plan (Figure 4.8). It is designed as a follow-up to the Galileo mission measurements, aiming to study the Jupiter system and its satellites in depth, with a focus on the largest moon, Ganymede. By thoroughly exploring the system and thereby unravelling its origin, its evolution and the formation of the different components such as the satellites, we will also get a handle on the history of the Solar System and the processes therein. The

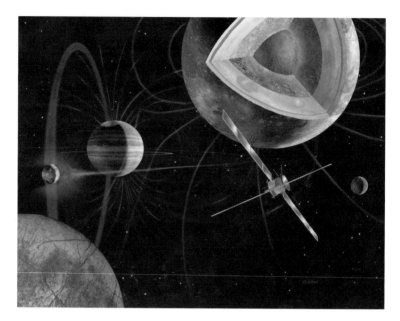

FIGURE 4.8 The JUICE spacecraft exploring the Jupiter system; artist's view. For colour version, see plates section. (Image credit: M. Carroll/ESA.)

overarching theme for JUICE as defined in the Yellow Book report is 'the emergence of habitable worlds around gas giants taking into account the necessary conditions involving the simultaneous presence of organic compounds, trace elements, water, energy sources and a relative stability of the environment over time'. The main questions that concern us in this book – the extent and evolution of habitable zones in the Solar System and the possible existence therein of habitable worlds as defined in Chapter 2 – are at the heart of the JUICE scientific goals for the Jovian system.

Ganymede is the focus of this mission for the reasons evoked above and because the features it shares with other celestial bodies make it the best example we have in our neighbourhood for studying habitable conditions in icy waterworlds in the Solar System and in other stellar systems. For Europa, the JUICE mission has two flybys planned above what are thought to be recently or currently active regions thinly covered with ice layers (regions called 'chaos') aiming to retrieve information on the underlying liquid ocean, its extent and composition, as well as on the morphology and chemical (water and other) composition of the surface. The investigations will focus on organic chemistry, its sources, sinks and evolution, and on identifying the best targets for landing in the future.

JUICE will explore the liquid-water oceans below the icy surfaces of the three moons Ganymede, Callisto and Europa looking for energy sources of chemical, thermal and tidal origin. In addition, the other habitability requirement, the stability of the environment on these moons, will be assessed by evaluating the gravitational coupling between the satellites and the planet. For the purpose of compared planetology and so as to be able to evaluate and appreciate the diversity among the Galilean moons, observations will also be made of Io (although it bears no interest for habitability) and other smaller moons.

The JUICE mission is expected to launch in mid-2022, for arrival at Jupiter in January 2030, after 7.6 years of interplanetary travel using an Earth–Venus–Earth–Earth gravity assist sequence, with a backup window of opportunity to launch in August 2023. The baseline for the mission's design is a spacecraft dry mass of 1900 kg and a propellant mass of 2900 kg, bringing the total launch mass up to 4.8 tonnes to be carried by an Ariane 5 vehicle. The total payload complement selected in 2013 has total mass of about 110 kg, and power requirements of 120–150 W depending on which instrument suites are operating. JUICE is a three-axis stabilized spacecraft whose power will be generated by a solar array of 60–70 square metre capable of producing between 640 and 700 W (Jupiter is probably as far as one can go on solar power with today's technology).

JUICE is foreseen to last a total of about 3.5 years. The current mission scenario is constructed in such a way as to allow JUICE to perform a tour of the Jovian system using gravity assists from the Galilean satellites to shape its trajectory. This tour will include continuous monitoring of Jupiter's magnetosphere and atmosphere, two targeted Europa flybys and a Callisto flyby phase reaching Jupiter latitudes of 30°, culminating with the dedicated Ganymede orbital phase. The current end-of-mission scenario involves spacecraft impact on Ganymede.

Owing to the difficulty of direct access to potential habitats within the Jovian moons, the *in situ* exploration of subsurface oceans is necessarily a long-term scientific endeavour. But on Ganymede and Europa, material (volatiles, minerals and perhaps organic species) rising from the internal ocean to the surface through fractures and cryomagmatic processes could be observed, revealing information on the deep liquid water.

Characterization of the ocean depth on Europa, its thickness and that of the boundary of its interface with the silicate mantle will be a task for a specifically designed Europa orbiter, studied in the framework of future missions to the Jovian system. Clearly, the characterization of the oceanic chemical and possibly prebiotic properties will be a longer-term scientific and technological challenge.

For technical and budgetary reasons, a surface lander or penetrator that could perform some *in situ* characterization of the surface ice of Ganymede and/or Europa is not possible for the moment. However, there are such projects currently being researched. The JUICE mission will nevertheless provide important information by establishing the existence of subsurface oceans under the crust of Ganymede, Europa or Callisto, defining their main characteristics and identifying privileged landing target sites.

## 4.3   SATURN'S SATELLITES

Further away from the Sun, at a distance of 10 AU, the Saturnian system (Figure 4.9) offers us not only the most popular of the giant planets (the 'Lord of the Rings') but also 62 natural satellites among which are several gems of possible habitats encased in rock, dust, ice and low temperatures. Among them are Titan and Enceladus, two moons of the outer Solar System that also show evidence of a subsurface water ocean from measurements performed by the Cassini–Huygens mission since 2004. Titan, the Earth-like moon with its rich organic chemistry, and Enceladus with its potential cryovolcanism and liquid subsurface water reservoir, are discoveries brought to us by space exploration which have revolutionized our perception of habitability in the Solar System.

FIGURE 4.9 A quintet of Saturn's moons come together in the Cassini spacecraft's field of view for this portrait. Janus (179 km across) is on the far left. Pandora (81 km across) orbits between the A ring and the thin F ring near the middle of the image. Brightly reflective Enceladus (504 km across) appears above the centre of the image. Saturn's second-largest moon, Rhea (1528 km across), is bisected by the right edge of the image. Rhea is closest to Cassini here. Mimas (396 km across) can be seen beyond Rhea, also on the right side of the image. The rings are beyond Rhea and Mimas, with Enceladus beyond the rings. (Image courtesy of NASA/JPL-Caltech/Space Science Institute.)

If we follow the water, then the discovery of water vapour plumes, issuing from fractures in the southern hemisphere of Enceladus, suggests the presence of a liquid water ocean (or pockets) in the interior, at distances from the Sun defying all previous notions of the habitable zone. There is ample evidence from Cassini–Huygens measurements that Titan has an internal liquid water ocean, while its rich Earth-like dense atmosphere has attracted interest since its discovery 75 years ago and is probably responsible for the only known case of (hydrocarbon) liquid seas exposed on a planetary surface. The surface of Enceladus is mostly water-ice, with a clear dichotomy between the smooth northern and the rugged southern hemispheres. From the south pole, through canyons known as 'tiger stripes' where the temperatures are warmer than in surrounding areas, strong plumes eject water vapour laden with salts, ammonia and organics out into space. If these geysers reflect the composition of the internal reservoir, then it is mostly composed of water with these compounds also incorporated. Different energy sources have been invoked for this small satellite's activity. The organic chemistry, the energy and the water which is obviously present on the satellite complete three out of the four requirements for habitability, but it is probably not a stable environment. On the other hand, Titan has an organic-rich atmosphere which deposits on its surface, but whether biological species also inhabit the internal water ocean remains unknown. They could hypothetically arise, though, by hydrolysis of organics in the chondritic matter that accreted to form Titan. A combination of Cassini–Huygens measurements with investigations made by future missions to the Kronian system is essential to help us understand the presence of liquid water at locations much more distant from the Sun than previously expected.

### 4.3.1  Titan: organic factory and habitat

Titan, Saturn's largest satellite, is another unique object in the Solar System (Figure 4.10). Although Titan is much colder than the Earth, as shown in Athena Coustenis and Frederic Taylor's books (*Titan: The Earth-like Moon* and *Titan: Exploring an Earth-like World*; Coustenis and Taylor, 1999 and 2008) the large satellite exhibits many similarities with the Earth and can be considered as a possible Class IV habitat. Recent Cassini–Huygens discoveries reveal that Titan is rich in organics from high atmospheric levels down to the surface, most probably contains a large liquid water ocean beneath its surface and has ample energy sources to allow for chemical evolution. Titan's atmosphere is composed of dinitrogen ($N_2$), like that of our own planet, with methane and hydrogen as the most abundant trace gases, and has a similar structure from the troposphere to the ionosphere, as well as an equivalent surface pressure of 1.5 bars – the only known extraterrestrial planetary atmospheric pressure close to that of the Earth. It has also been demonstrated that the sources for nitrogen on Earth and Titan are similar. The stratosphere and mesosphere together form what we call the middle atmosphere. The next atmospheric layer, the thermosphere (which on Earth covers the altitudes from 80 to 500 km), together with the exosphere forms the upper atmosphere. The ionized component of the thermosphere is the ionosphere, which is co-localized and interacts with the thermosphere. Titan's upper atmosphere is not protected (as the Earth's is) by an intrinsic magnetic field, and therefore it is subject to a

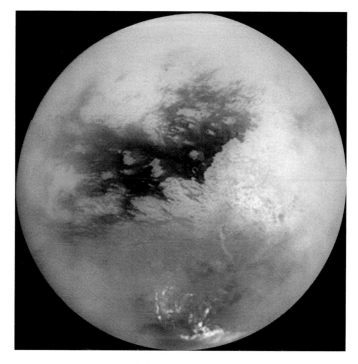

FIGURE 4.10  A mosaic of nine processed images acquired during
Cassini's first very close flyby of Saturn's moon Titan on 26 October 2004,
giving a detailed full-disk view of the moon. The view is centred on 15°
south latitude, and 156° west longitude. Brightness variations across
the surface and bright clouds near the south pole are exposed. The
images in this mosaic were acquired from distances ranging from
650 000 km to 300 000 km. (Image courtesy of NASA/JPL/Space Science
Institute.)

direct bombardment by energetic electrons, protons and oxygen ions. The
ionosphere is formed through ionization by the solar radiation and by
electron/ion impacts on the neutral atoms and molecules (electron strip-
ping and charge exchange interactions). On Titan it is the place where
complex ion–molecule reactions take place leading to the formation of
hydrocarbons and nitriles, as studied by the Cassini Ion and Neutral Mass
Spectrometer (INMS), and is found to extend up to ~1500 km. Right above
this is the exosphere, where collisions between particles are rare and
where the dominant force is the gravitational pull, which in the case of

Titan extends its influence (and thus provides an estimate for the limit of its exosphere) to about 50 000 km. The exosphere is thus the region from where atmospheric particles can eventually escape into space. Charged energetic ions from Saturn's magnetosphere can interact with neutral atoms from Titan's exosphere and become energetic neutral atoms (ENAs), as discussed by Iannis Dandouras and colleagues (2009).

Methane on Titan seems to play a role comparable to that of water on the Earth, with a complex cycle that has yet to be fully understood (Figure 4.11). Methane can exist as a gas, liquid and solid, since the mean surface temperature is almost 94 K as measured by the Atmospheric Structure Instrument on board the Huygens probe, approaching the triple point of methane. However, the amount of methane currently found in Titan's atmosphere is a mystery, since methane photolysis and recombination would have reduced these amounts to nothing a long time ago. To explain this paradox, methane reservoirs on or under the surface have been hypothesized. Analogies

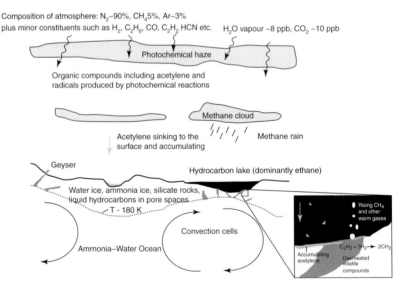

FIGURE 4.11 Methane cycle, environmental meteorology and biology in Titan's atmosphere. (Adapted from F. Bagenal's LASP class, http://lasp. colorado.edu/~bagenal/3720/CLASS23/23Titan.html. Sketch by D.Grinspoon.)

can also be made between the current organic chemistry on Titan and the prebiotic chemistry which was active on the primitive Earth. In spite of the absence of permanent bodies of liquid water on Titan's surface, the chemistry is quite similar.

### 4.3.1.1 Titan's organic chemistry throughout the atmosphere

Our previous understanding of Titan's chemistry was constrained to the stratospheric region, basically between 100 and 500 km, because that was where previous Voyager measurements probed the atmosphere. But with Cassini and its Ion and Neutral Mass Spectrometer (INMS), a major discovery was made: the haze surrounding the satellite forms more than 1000 km above the surface, through a combination of ion and neutral chemistry initiated by energetic photon and particle bombardment of the upper atmosphere. This energetic chemistry in the ionosphere produces large molecules such as benzene, which condense out to create the haze we see today on Titan (Figure 4.12). The thick haze layer where the organics detected on Titan precipitate could be an analogue to the UV-protective smog that sheltered the early Earth. As the haze particles fall through the atmosphere, they accrete and increase in size, becoming large polymers. Measurements throughout the atmosphere have indicated the presence of numerous hydrocarbon and nitrile gases, as well as a complex layering of organic aerosols (tholins) that persists all the way down to the surface (Figure 4.12). In the presence of nitrogen and methane, Titan's atmosphere is among the most appropriate environments known for pre-biotic synthesis, and indeed, several of the organic compounds we find on Titan today, such as hydrogen cyanide (HCN), cyanoacetylene ($HC_3N$) and cyanogen ($C_2N_2$), were major players in the Earth's prebiotic chemistry. In particular, the presence of benzene is extremely interesting, as it is the only polycyclic aromatic hydrocarbon (PAH) discovered on Titan today. The presence of PAHs on Titan's atmosphere is important as they may contribute to the synthesis of biological building blocks. Moreover, the combination of the liquid deposits on the surface of Titan and the low temperature could create the proper environment for this biosynthesis.

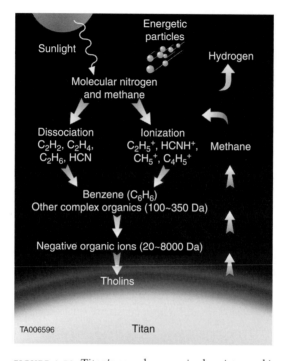

FIGURE 4.12   Titan's complex organic chemistry and its structure with
the atmospheric layers as updated after the discoveries made by the
Cassini–Huygens mission INMS, UVIS, CIRS and GCMS instruments. For
colour version, see plates section. (Image courtesy of NASA/JPL/H. Waite.)

Recent laboratory experiments have shown that aromatic compounds
have a good chance of being produced on icy surfaces.

The large molecules and aerosols produced on Titan strongly
absorb solar and visible radiation and are thus essential in heating
Titan's stratosphere (between 100 and 500 km) and forming wind systems in the middle atmosphere, a situation similar to ozone's contribution in the Earth's middle atmosphere. The large organic inventory of
Titan's atmosphere is eventually deposited on Titan's surface, forming
dunes and other deposits. Ninety per cent of the energy at the surface of
Titan is held in by a greenhouse effect due to nitrogen, methane and
hydrogen. These are symmetrical molecules which normally do not
have a greenhouse effect on Earth, but they do on Titan, owing to the

FIGURE I.I The Earth from space. The presence of water is detectable remotely from the dark blue colour of the oceans, the water vapour clouds and the northern polar cap. (Image courtesy of NASA/GSFC/Suomi NPP.)

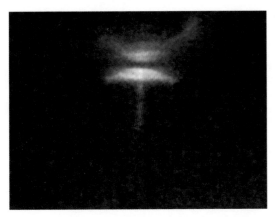

FIGURE I.2 A protoplanetary disk, HH-30 in Taurus, about 450 light years away, observed with the Hubble Space Telescope. The disk emits a stellar jet, aligned with the rotational axis of the disk which appears in black as an absorption in front of the stellar light. (Image courtesy of STSci/ESA/ NASA.)

FIGURE I.4 A schematic view of the Solar System (not to scale), with the terrestrial planets close to the Sun and the giant planets at further distances, beyond the main asteroid belt. A comet is shown near the orbits of Jupiter and Saturn. (Image courtesy of NASA/JPL.)

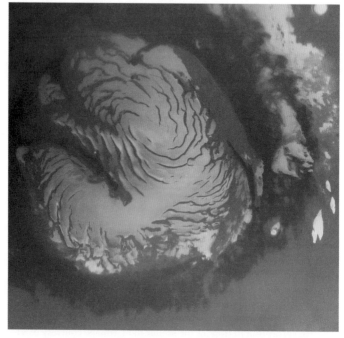

FIGURE I.7 Water-ice on the Northern perennial polar cap of Mars. The image was taken with the camera of the Mars Global Surveyor spacecraft in 1999. (Image courtesy of NASA/JPL.)

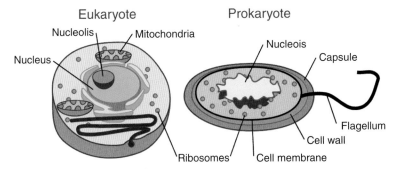

FIGURE 2.4 Cell types: (left) eukaryote, (right) prokaryote. (Image from the US Science Primer, a work of the National Center for Biotechnology Information, part of the National Institutes of Health.)

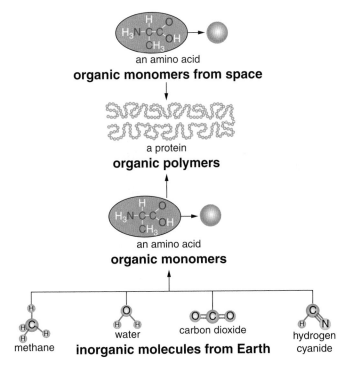

FIGURE 2.6 Prebiotic Earth with the sources for organic chemistry from exogenous and endogenous inputs. (Image by Jen Philpot and Jane Wang, courtesy of *Science Creative Quarterly*.)

FIGURE 2.7 Liquid water seas on Earth; artist perspective by T. Encrenaz.

FIGURE 2.11 Río Tinto, a river with extreme acidic living conditions where some Earth organisms still survive, thought to have similarities with some places of subterranean Mars. (Image credit: J. Segura and R. Amils.)

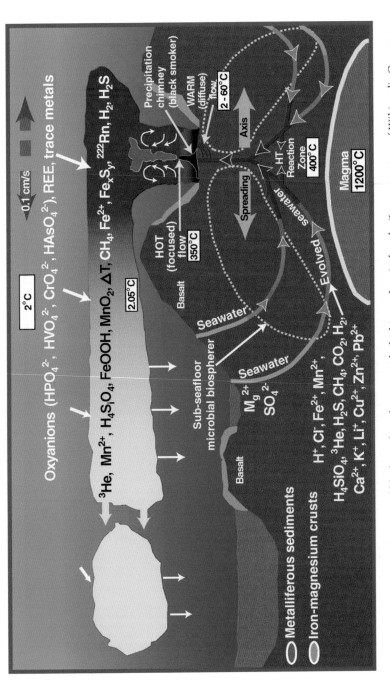

FIGURE 2.10 Hydrothermal vents and black smokers diagram with the biogeochemical cycle. [Image courtesy of Wikimedia Commons, http://en.wikipedia.org/wiki/File:Deep_sea_vent_chemistry_diagram.jpg.]

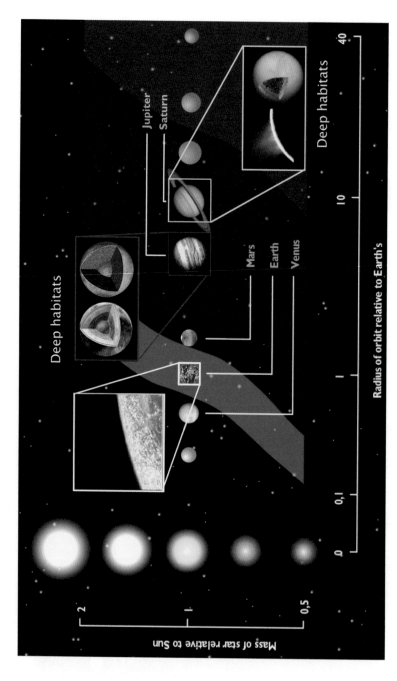

FIGURE 2.12 The habitable zone in our Solar System and elsewhere. (Adapted from Lammer *et al.*, 2009.)

FIGURE 2.14 The Horsehead Nebula. The dark region corresponds to dense molecular clouds where interstellar chemistry is active. (Image courtesy of Adam Block, Mt Lemmon SkyCenter, University of Arizona.)

FIGURE 3.3 The three terrestrial planets with an atmosphere, showing their relative sizes. Mars (left) is the smallest; its surface is covered with deserts and the polar caps are visible; the Earth (centre) shows a large-scale cloud system of water-ice; Venus (right) is covered with a thick cloud of sulfuric acid. (Image courtesy of NASA.)

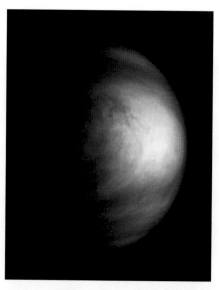

FIGURE 3.4 Planet Venus, observed in the ultraviolet by the camera of the Galileo probe during its flyby of the planet in February 1990. The UV radiation probes the top of the thick cloud layer, mostly composed of sulfuric acid. (Image courtesy of Galileo Project/JPL/NASA.)

FIGURE 3.11 Mosaic of the Valles Marineris hemisphere of Mars projected into point perspective, a view similar to that which one would see from a spacecraft. The viewer's distance is 2500 km from the surface of the planet. The mosaic is composed of 100 Viking Orbiter images of Mars. The centre of the scene (latitude −7°, longitude 78°) shows the entire Valles Marineris canyon system, over 3000 m long and up to 8 km deep, extending from Noctis Labyrinthus, to the west, to the chaotic terrain to the east. The three Tharsis volcanoes (dark spots), each about 25 km high, are visible to the west. (Image courtesy of NASA/MDIM.)

**WATER MAP**
2001 Mars Odyssey Gamma Ray Spectrometer
H₂O Low ▭ H₂O High

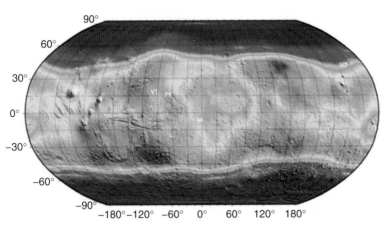

FIGURE 3.12 Water under the surface of Mars, as detected by the Gamma Ray Spectrometer of the Odyssey mission in 2000. The instrument measured the nuclear radiation emitted by the surface, following the impact of cosmic rays. The abundance of neutrons emitted by the surface is an indicator of the content of hydrogen atoms below the surface. A low content of neutrons indicates high hydrogen abundance and thus a high water content. In each hemisphere, the map was obtained during summer, to avoid the presence of the carbon dioxide seasonal cap. (Image courtesy of NASA/JPL.)

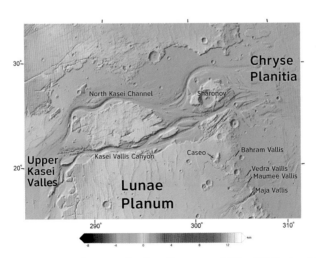

FIGURE 3.13 An example of a valley network (Kasei Valles) on Mars, near Chryse Planitia, indicating the presence of liquid water in the past history of the planet. (Image courtesy of NASA/Google Earth, via Wikimedia Commons.)

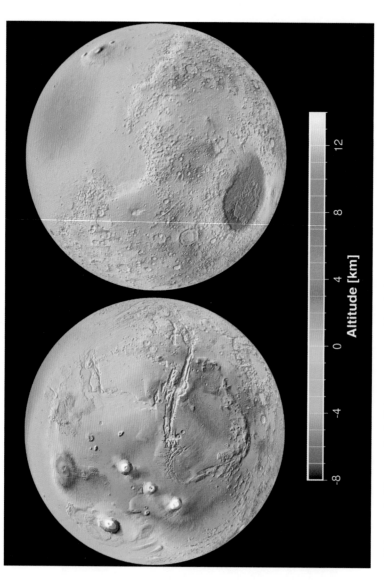

FIGURE 3.14 Altimetry (height mapping) of Mars, as measured by the laser altimeter experiment (MOLA) of the Mars Global Surveyor mission. The white dots over the red area on the left image are the Tharsis volcanoes. The dark blue spot on the right image is the bottom of Hellas basin, a giant impact crater. (Image courtesy of NASA/JPL.)

FIGURE 3.15 A self-portrait of Curiosity, the rover of the Mars Science Laboratory mission, from a combination of several high-resolution images taken by the Mars Hand Lens Imager. The rover has been in operation on the surface of Mars since August 2012, when it landed in Crater Gale. The robotic arm is designed to collect samples at the surface and below. The samples are analysed by the rover instruments for chemical and mineralogical characterization. (Image courtesy of NASA/JPL-Caltech/ Malin Space Science Systems.)

FIGURE 3.17 Oceans on Earth: the origin of water on our planet is a pending question. (Image courtesy of NASA.)

FIGURE 3.18 Earth and Moon system as observed by the Galileo spacecraft during its December 1992 flyby. (Image courtesy of NASA.)

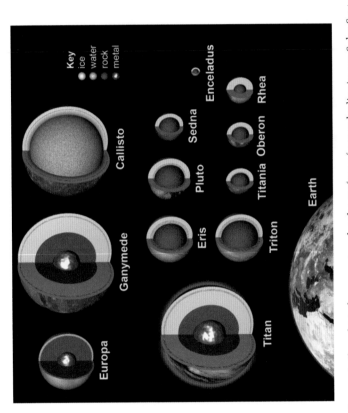

FIGURE 4.1 Liquid water is thought to be present under the surface of many bodies in our Solar System, particularly the Galilean moons of Jupiter, such as Europa, Callisto and Ganymede, and Saturn's Titan and Enceladus. In addition, models of heat retention and heating via radioactive decay in smaller icy bodies suggest that Saturn's Rhea, Uranus' Titania and Oberon, Neptune's Triton, trans-Neptunian objects and dwarf planets Pluto, Eris, Sedna and Orcus may also have oceans underneath solid icy crusts approximately 100 km thick. The models vary and in some cases predict that the liquid layers are in direct contact with the rocky core, which allows efficient mixing of minerals and salts into the water, while in others layers of high-pressure phases of ice are thought to underlie the liquid water layer. (Image credit: Doug Ellison, for the Planetary Society).

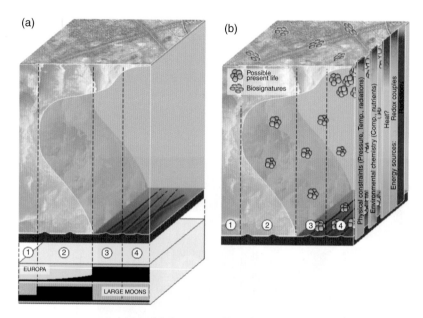

FIGURE 4.2 (a) Possible locations of liquid layers in the icy moons of Jupiter are plotted here as a function of depth (b) Current habitability of Europa. Possible locations of present life and biosignatures have been plotted as a function of depth. For full caption, see p. 127 in text.

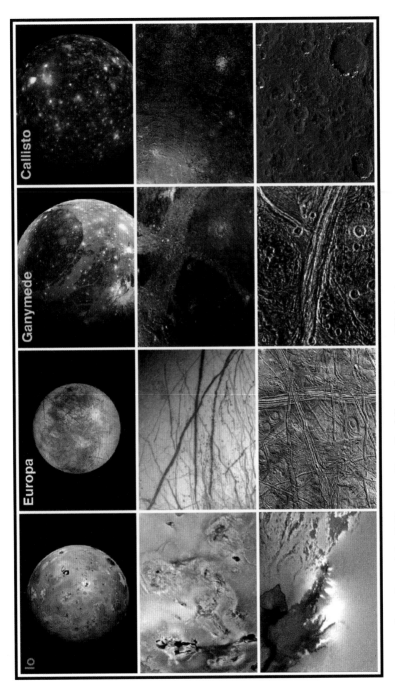

FIGURE 4.3 The surfaces of the Galilean satellites. For full caption, see p. 132 intext.

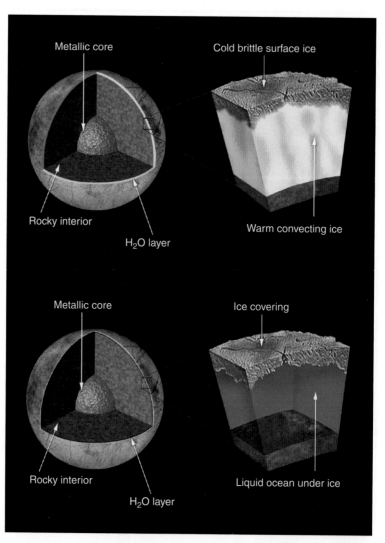

FIGURE 4.5 These artist's drawings depict two proposed models of the subsurface structure of the Jovian moon, Europa. Galileo's observations of surface features can be explained in two ways: either by the existence of a warm, convecting ice layer, located several kilometres below a cold, brittle surface ice crust (top), or by a layer of liquid water with a possible depth of more than 100 km (bottom). (Image courtesy of NASA/JPL.)

FIGURE 4.8 The JUICE spacecraft exploring the Jupiter system; artist's view. (Image credit: M. Carroll/ESA.)

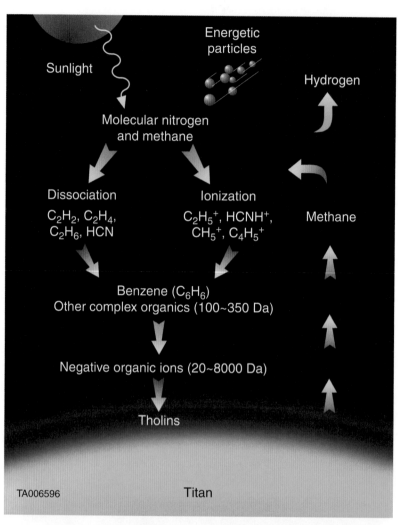

FIGURE 4.12 Titan's complex organic chemistry and its structure with the atmospheric layers as updated after the discoveries made by the Cassini–Huygens mission INMS, UVIS, CIRS and GCMS instruments. (Image courtesy of NASA/JPL/H. Waite.)

FIGURE 4.13A Hydrocarbons on Titan. The dark spots are liquid hydrocarbon lakes detected in Titan's northern hemisphere.(Image courtesy of NASA/JPL-Caltech/USGS.)

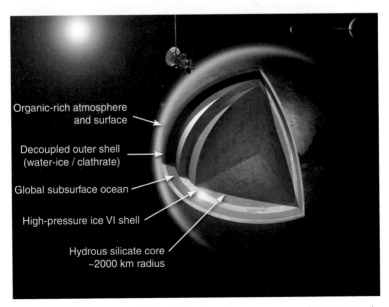

Organic-rich atmosphere and surface

Decoupled outer shell (water-ice / clathrate)

Global subsurface ocean

High-pressure ice VI shell

Hydrous silicate core ~2000 km radius

FIGURE 4.14 Layers of Titan's interior. (Image courtesy of A. D. Fortes/ UCL/STFC, via NASA.)

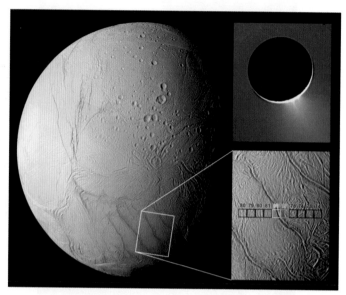

FIGURE 4.15 Surface of Enceladus, with focus on tiger stripes and an optical image of the jets emerging therefrom. The right lower zoom indicates the temperature scale across a south polar region, with the warmest temperatures detected at the 'tiger stripe' location (in yellow). (Image courtesy of NASA/JPL/Space Science Institute.)

FIGURE 4.17 The Titan Saturn System Mission concept with the dedicated orbiter, the lake lander and the floating hot-air balloon; artist's view. (Image by C. Waste, courtesy of NASA/JPL.)

FIGURE 4.18 Comet Halley as shown on the Bayeux tapestry relating the conquest of England by William the Conqueror (1066). (Image courtesy of Wikipedia, http://en.wikipedia.org/wiki/Halley's_Comet.)

FIGURE 4.19 Comet P/Halley as taken on 8 March 1986 by W. Liller, Easter Island, part of the International Halley Watch (IHW) Large Scale Phenomena Network. (Image courtesy of NSSDC/NASA.)

FIGURE 4.27 The surface of Neptune's largest moon, Triton, at its sub-Neptunian hemisphere. This image is a false-colour mosaic taken in 1989 by NASA's Voyager 2 spacecraft. For full caption, see p. 181 in text.

FIGURE 4.28 Artist's view of Triton's geysers. Triton is scarred by enormous cracks. Voyager 2 images showed active geyser-like eruptions spewing nitrogen gas and dark dust particles several kilometres into the atmosphere. (Image © Ron Miller, courtesy of International Space Art Network, http://spaceart1.ning.com/photo/triton-geyser.)

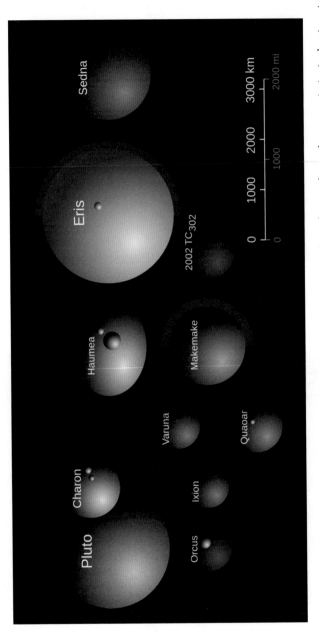

FIGURE 4.29 Some of the largest trans-Neptunian objects. The arcs around Makemake and Eris indicate the uncertainties in the size, given the unknown albedo. (Image courtesy of Wikimedia Commons.)

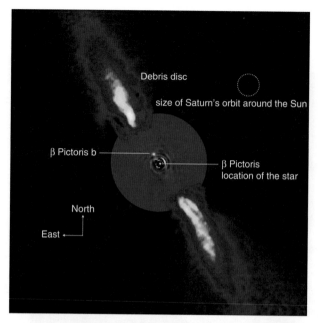

Debris disc

size of Saturn's orbit around the Sun

β Pictoris b

β Pictoris
location of the star

North

East

FIGURE 5.5 The disk of Beta Pictoris, first detected with the telescope of Las Campanas. This was the first identification of a debris disk around a young star. In 2009, a planet was discovered close to the star, orbiting in the plane of the debris disk. (Image © ESO/A.-M. Lagrange *et al.*)

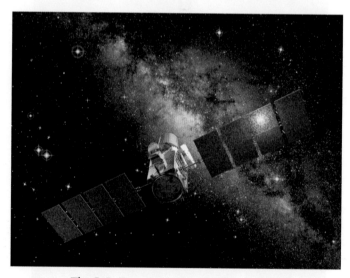

FIGURE 5.7 The CoRoT space mission. CoRoT was launched by CNES into Earth orbit in December 2006 with the objective of detecting exoplanets in transit. After 6 years of operation, CoRoT has detected about 25 exoplanets, and many more are waiting confirmation. (Image © CNES/ D. Ducros.)

FIGURE 5.8 The Kepler space mission. Launched in March 2009, the Kepler satellite is designed for detecting exoplanets by transit. Kepler has detected about 100 confirmed exoplanets and more than 2000 possible candidates. (Image courtesy of NASA/Kepler mission/Wendy Stenzel.)

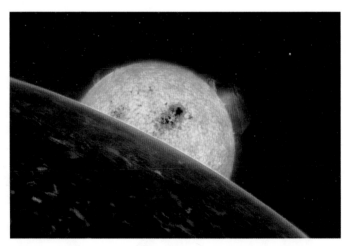

FIGURE 5.12 The exoplanet CoRoT-7 b, an extreme example of a small exoplanet located in the immediate vicinity of its host star; artist's impression. (Image © ESO/ L. Calçada.)

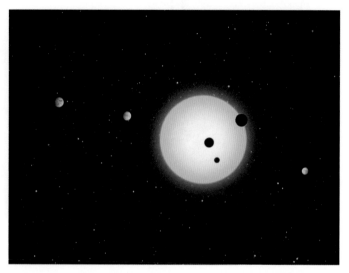

FIGURE 5.13 The Kepler 11 planetary system with six planets around their star, discovered by the satellite in 2011; artist's impression. (Image courtesy of NASA/T. Pyle.)

FIGURE 5.14 Artist's impression of an exoplanet orbiting both stars in a binary star system. (Image credit: original art by T. Encrenaz.)

FIGURE 5.15 Artist's impression of an Earth-like world in a different planetary system. (Image credit: original art by R. Poux.)

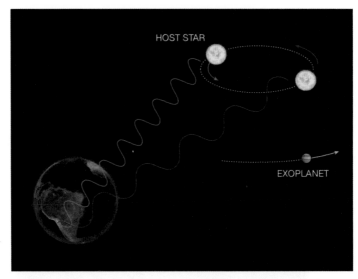

BOX 5.1 The velocimetry method to detect exoplanets. The radial velocity of the star is measured as it is modulated by the star's motion with respect to the centre of gravity of the star-planet system. As the star moves toward the Earth, the signal emitted by the star is blue-shifted, i.e. moved towards higher frequencies (or shorter wavelengths); as it moves back in the other direction, the signal is red-shifted, i.e. moved towards lower frequencies (longer wavelengths). The velocity method allows astronomers nowadays to detect radial velocities lower than $1~\mathrm{m\,s^{-1}}$, bringing the domain of Earth-like planets within detectability limits. (Image © ESO.)

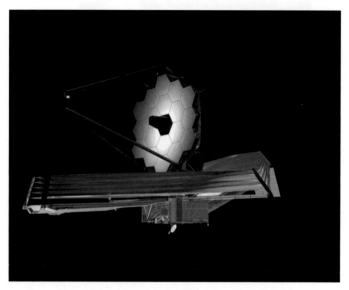

FIGURE 6.2 The James Webb Space Telescope project. (Image courtesy of NASA.)

FIGURE 6.3 Potential design for ATLAST, a 16-m, segmented-mirror future telescope. (Image by Bill Oregerle (NASA/GSFC and Marc Postman (STScI), courtesy of NASA/STScI.)

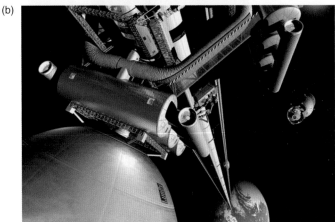

FIGURE 6.8 (a) The Bernal Sphere is a concept for hosting large numbers of humans upon the interior surface of a large rotating sphere, about 1 km in circumference, which would use centrifugal force as a substitute for gravity. (Image courtesy of NASA.) (b) The space elevator concept consists of a cable or tether fixed to the Earth's equator, reaching into space to facilitate space travel. (Image by Pat Rawling, courtesy of NASA/MSFC.)

FIGURE 6.9 Images corresponding to an imaginary Mars transition towards an Earth-like planet. (Image by D. Ballard, courtesy of Wikimedia Commons.)

FIGURE 6.10 The Arecibo radiotelescope. The image shows the bridge allowing access to the focal laboratory suspended above the telescope's centre. (Image courtesy of the NAIC – Arecibo Observatory, a facility of the NSF.)

FIGURE 6.11 The SETI Project: the Allen Telescope Array's first phase of 42 telescopes as completed. To build the full observatory (350 antennas), the SETI Institute will need additional funding. (Image © Zack Frank/ Shutterstock.com)

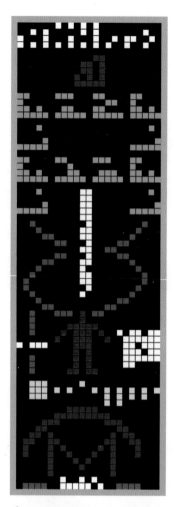

FIGURE 6.12 The Arecibo message as sent in 1974 from the Arecibo Observatory. From bottom to top: the Arecibo telescope; a picture of the Solar System; a human figure; the double DNA helix; the formulas of the sugars and bases in the nucleotides of DNA; the atomic numbers of hydrogen, carbon, nitrogen, oxygen and phosphorus, which make up deoxyribonucleic acid (DNA); and the numbers 1 to 10. (Image by Arne Nordmann, courtesy of Wikimedia Commons.)

dense atmosphere. More interesting still, Titan is the only world in the Solar System to have an 'anti-greenhouse' effect, caused by the haze layers in the atmosphere, that lets light in and stops infrared. This anti-greenhouse effect is half as strong as the greenhouse effect. The tropospheric emission temperature (near the tropopause, at 40 km altitude) is determined by the anti-greenhouse effect and is 9 K cooler than the effective temperature (that is, the temperature equivalent to a black body emitting the same total amount of electromagnetic energy). The increase in temperature of 21 K from the tropopause to the surface is due to a greenhouse effect resulting from thermal infrared radiation emitted from the lower atmosphere and warming the surface. The surface is not in radiative balance, because convective motions account for an energy flux of 1 per cent.

Hence, on Titan, the upper thermosphere is linked intimately with the middle atmosphere and even the surface. But more than that, it is possible that the scheme we have found to be operating for the production of organic molecules on Titan is also the explanation for the production and delivery of prebiotical components to the early terrestrial oceans, much as the Miller–Urey experiment hypothesized.

Furthermore, the Gas Chromatograph and Mass Spectrometer (GCMS) on board the Huygens probe, which successfully landed on Titan's surface on 14 January 2005, did not detect a large variety of organic compounds in the lower atmosphere. In a 2006 article, François Raulin and colleagues indicated that 'the mass spectra collected during the descent show that the medium-altitude and low stratosphere as well as the troposphere are poor in volatile organic species, with the exception of methane' (Raulin *et al.*, 2006). Condensation of these species on aerosol particles is a probable explanation for the absence of the volatiles in the measurements. They were captured and analysed by the Huygens Aerosol Collector and Pyrolyzer (ACP) instrument on board the Huygens probe, which found them to be made of refractory organic nuclei covered with the volatile compounds that would have condensed on them. During pyrolysis, ammonia and hydrogen cyanide were released, supporting the tholin hypothesis, but there is still a need to measure the abundances of the condensates and their elemental composition, and to determine their molecular structure.

The measurements by the Cassini instruments of gases and aerosols, and in particular tholin material, are supported by laboratory simulation experiments that try to reproduce the organic chemistry detected on Titan, integrating the information available on the energy sources and the processes as well as possible. Several teams, led by experts in this field such as Bishun Khare, François Raulin and Hiroshi Imanaka, have conducted such experiments on the chemical evolution of $N_2$–$CH_4$ mixtures on Titan. The results, which are quite representative of what is found on Titan, tend to indicate that this work is essential since it manages to mimic the real chemical processes in Titan's atmosphere.

Titan is thus the largest abiotic organic factory in the Solar System. Indeed, as estimated and demonstrated by Ralph Lorenz, the quantities of methane and its organic byproducts in Titan's atmosphere, seas and dunes exceed the carbon inventory in the Earth's ocean, biosphere and fossil fuel reservoirs by more than an order of magnitude (Lorenz, 2008). The degree of complexity that can be reached from organic chemistry in the absence of permanent liquid water bodies on Titan's surface is still unknown, but it could be high.

Moreover, Titan is the only planetary object, besides Earth, with long-lived, exposed bodies of liquid on its surface (Figure 4.13). The features range in size from less than 10 $km^2$ to at least 100000 $km^2$. They are limited to the region poleward of 55° N. By 2009, Cassini's instruments had identified and mapped almost 655 geological structures referred as lakes and/or basins, mostly in the northern polar region. Large cloud systems, some of which attain the size of terrestrial hurricanes (1000 km across), appear occasionally, while smaller, transient and temporary features exist on a daily basis above these lakes and also can be found at mid-latitudes. It has been theorized that the lakes on Titan could be the result of condensation and even rain processes in Titan's atmosphere above. The hydrocarbon cycle

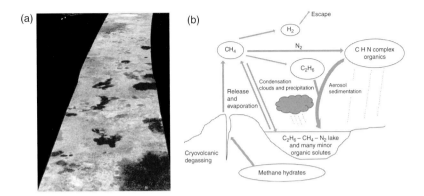

FIGURE 4.13 Hydrocarbons on Titan. (a) The dark spots are liquid hydrocarbon lakes detected in Titan's northern hemisphere. For colour version, see plates section. (Image courtesy of NASA/JPL-Caltech/USGS.) (b) Methane (CH$_4$) is released into the atmosphere from Titan's interior stores through volcanic action, and evaporates from the lakes of methane and ethane (C$_2$H$_6$) identified by the Cassini spacecraft on the satellite's surface. Chemical reactions in the atmosphere convert it to ethane, complex organic aerosols and hydrogen gas (H$_2$), which escapes into space. Ethane and methane partly condense, forming clouds and hazes that rain out, replenishing the lakes and depositing many organic species in solution. (Image © F. Raulin, LISA, Université Paris Est-Créteil Val de Marne, France.)

involving methane and ethane causes the liquid to be re-injected into the atmosphere where it rains out again after some time, producing the seas but also the fluvial features observed by Huygens near the equator.

In spite of the low temperature prevailing on Titan, we are not talking about a replica of a frozen Earth: the chemical system is evolving, and it may well produce compounds of (astro)biological interest, in particular on the surface where the organics are deposited after their descent through the atmosphere. Some experiments have even argued for the creation of amino acids through the reaction of the organics on the surface with any possible liquid water or even water-ice. Those processes could be particularly favourable in zones of Titan's surface where cryovolcanism might be occurring, or deeper in the interior where hydrothermal vents, similar to Earth's black smokers, may exist, albeit in different conditions, as has been suggested in recent research. Even with the detection of the large lakes in the north, Cassini was unable to detect any viable source that could re-supply the total amount of methane currently found into the atmosphere. Cryovolcanic outgassing has been hypothesized to re-inject methane into the atmosphere, yet over what timescales and through which internal processes remains unknown, even though several areas are currently believed to have been formed under the influence of cryovolcanism.

Cryovolcanism is considered to be one of the principal geological processes that has shaped several of the icy moons' surfaces. This activity can be described as ice-rich volcanism, while the cryovolcanic ejecta are referred to as cryomagma. The cryomagma might appear in the form of icy liquid and, in some cases, partially crystallized slurry, with unknown precise composition. There are currently three major cryovolcanic candidate regions, all of which are located close to Titan's equator. Moreover, modelling of Cassini/VIMS data taken from these regions from several flybys is providing signs of albedo variations with time, suggesting possible fluctuating deposition of material from the interior and from the atmosphere. If further analysis confirms this suggestion, then cryovolcanism will undoubtedly be considered as one of the top processes that replenish methane in the atmosphere and reshape and change the surface.

Cassini–Huygens also found that the inventory of geological processes shaping the surface – aeolian, fluvial, impacts, tectonics, cryovolcanism, etc. – is broadly similar to the Earth's (called 'morphotectonics'), more so than for Venus or Mars. This makes Titan our best analogue so far to an active terrestrial planet, albeit with different materials and physical conditions. Cosmic rays reaching Titan's surface could well irradiate any liquid bodies present there, thus assisting organic syntheses. In addition, the interface between the liquid phase and the solid icy deposits on the ground may include sites of catalytic activity favourable to chemical reactions. Titan's lakes present an ideal place to look for such effects.

### 4.3.1.2 Methane-based biology and the lakes

Earth-like life needs liquid water. This is currently impossible on Titan's surface, given the low temperatures, but may have been possible in the past, through impacts which could have created ephemeral pools of water in their craters from melting ice, temporarily allowing the emergence or growth of life. In addition, although the chemical reactions that lead to life on Earth need liquid water as a solvent, they take place almost entirely between organics. In the search for life and habitable conditions in the Solar System, one should therefore not ignore the study of organic chemistry, and Titan is a good place to start.

Titan's organic inventory has been known for quite some time now, ever since the discovery of hydrocarbons in its atmosphere by Kuiper, Gillett and others in the middle of the twentieth century. We now know, thanks to Cassini, that in addition to the atmosphere, organics are also widespread across the surface in different phases and in the lakes, seas, dunes and channels. Thus, all the ingredients that are supposed to be necessary for life to appear and even develop – liquid water, organic matter and energy – seem to be present on Titan.

Consequently, it cannot be ruled out that life may have emerged on or in Titan at some point in its history, in spite of the extreme and inhospitable conditions of its current environment. As we have seen in Chapter 2, life may have been able to adapt and to persist for some time even if the conditions (pH, temperature, pressure, salt concentrations) are not compatible with life as we know it. However, the detection of any potential biological activity in Titan's current internal water ocean seems very challenging – more so when we note that besides Earth-like lifeforms, other possible forms of living organisms have been speculated to exist on Titan.

Chris McKay and Heather Smith in 2005 noted the astrobiological importance of the liquid hydrocarbon lakes on Titan and hypothesized that a lifeform called 'methanogens' might consume hydrogen instead of oxygen, a hypothesis that could be tested against measurements in the lower atmosphere. In other research work by Darrell

Strobel and co-workers in 2010, based on data from the Cassini orbiter focusing on the complex chemical activity on the surface of Titan, hydrogen was shown to be precipitating through Titan's atmosphere and then mysteriously disappearing on the surface in a fashion similar to oxygen consumption on Earth. If this effect was produced by a living organism on Titan it would have to be markedly different from an Earth-like organism, but even so it has attracted interest as a hypothetical second form of life without the need for water. However, in more recent work it has been shown that the atmospheric haze can also absorb and desorb the hydrogen, thus providing the missing sink without any requirement for life.

Another measurement result that has been interpreted as a possible indicator for some sort of lifeform existing on Titan is the lack of acetylene on the surface: there is no clear evidence so far of this compound in the data received from Cassini, although it is expected to have been deposited through the atmosphere. Roger Clark and colleagues (2010) have suggested that this could be because some living organism on the surface is using acetylene as an energy source. This theory is much debated and controversial among the scientific community, especially because the phenomenon could be of non-biological origin, but it has the merit of inspiring new and interesting astrobiological theories. According to one theory put forth by astrobiologists, a hypothesized 'methane-based life' would consume not only methane but also hydrogen. However, another possibility, formulated by Mark Allen of JPL (2010), is that 'sunlight or cosmic rays are transforming the acetylene in icy aerosols in the atmosphere into more complex molecules that would fall to the ground with no acetylene signature'.

To date, methane-based lifeforms are hypothetical; they have not been detected anywhere in our Solar System, although we do find 'methanogens' on Earth, which are liquid-water-based microbes that feed on methane or produce it as waste. At Titan's low temperatures, and in the absence of any liquid water, which as we have seen would be frozen on the surface, a methane-based organism would need to resort to liquid methane or its byproducts such as ethane. However, as we have seen, liquid water is not a strict requirement, nor does it have to be on the surface. In Titan's putative ocean, organics penetrating through the icy crust might find the liquid water and produce a different methane-based lifeform. But with current technological means and plans, we are not even close to proving its existence.

### 4.3.1.3 Subsurface ocean on Titan

A combination of different land-shaping processes, such as aeolian, fluvial, and possibly tectonic and endogenous cryovolcanic processes, operates on Titan. Linear features and possible cryovolcanic spots are found, in general, close to the equator. In particular, elevated as well as fractured crustal features are observed, and the fact that these features are locally regrouped indicates a morphotectonic pattern. Their shapes, sizes and morphologies suggest that they are tectonic in origin, although it may be a different form of tectonism from the terrestrial one, originating from internal compressional and/or extensional activities. The triggering mechanism that leads to such dynamic movements is possibly Saturn's tidal pull, whose effects concentrate around the equator.

Titan is tidally locked with respect to Saturn and thereby subject to periodic tidal forcing of its interior and surface. The recent detection of periodic tidal stresses on Titan, caused by Saturn's gravity as the satellite revolves around the planet, shows deformations that are larger (perhaps as big as 10 m) than would be expected in a purely solid rocky body, and this may be consistent with a global ocean at depth (Figure 4.14).

In addition, the presence of an internal liquid water ocean on Titan is supported by models based on radar and gravity Cassini measurements and those from the Atmospheric Science Instrument on the Huygens probe (HASI). Indeed, the extremely low-frequency electric signal recorded by HASI measurements was recently interpreted as a Schumann resonance between Titan's ionosphere and a conducting ocean, probably of small dimensions and located at some 50 km under the surface. A Schumann resonance is manifested by the presence of extremely low frequency (ELF) radio waves, detected, in the case of Titan, in electric conductivity measurements acquired by the Huygens

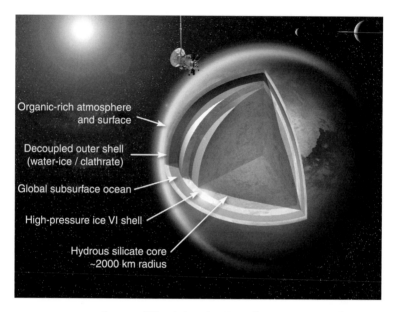

Organic-rich atmosphere
and surface

Decoupled outer shell
(water-ice / clathrate)

Global subsurface ocean

High-pressure ice VI shell

Hydrous silicate core
~2000 km radius

FIGURE 4.14 Layers of Titan's interior. For colour version, see plates section. (Image courtesy of A. D. Fortes/UCL/STFC, via NASA.)

probe during the descent. Such waves are also found on Earth where they are reflected by both the surface and the ionosphere in a configuration where the Earth's atmosphere resembles a giant 'sound box' where certain frequencies of ELF waves resonate and become stronger, while others die away. On Titan, however, the surface is a poor reflector because of its low conductivity and so these waves penetrate the interior where they may have been reflected by the liquid–ice boundary of a subsurface ocean of water and ammonia. Thermal evolution models of Titan by Ralph Lorenz and colleagues concur with this interpretation, suggesting that the moon may have an ice crust between 50 and 150 km thick, lying atop a liquid water ocean a couple of hundred kilometres deep. On Titan, unlike Ganymede or Callisto, the presence of nitrogen in the atmosphere suggests that ammonia could be found in the interior where, as we have seen above, it could act as an antifreeze. Beneath this one would find a layer of high-pressure ice.

In 2012, Luciano Iess and co-workers suggested the existence of a globe-encircling, shallow liquid water ocean as the most probable interpretation of Cassini's measurements of the tidal contributions to the non-spherical part of Titan's gravity field (Iess *et al.*, 2012). Indeed, the determination of the tidal potential Love number (which defines the capability for a rigid body to deform under tidal stresses) from Cassini gravity measurements indicates a fluid response on a tidal timescale. In addition, Cassini's measurement of asynchronicity in Titan's rotation can be interpreted to be a result of decoupling the crust from the deeper interior through the presence of a liquid layer.

Although we have some observational evidence today for suspecting the presence of a liquid water–ammonia ocean on Titan, more measurements are needed. In particular, precise knowledge of tidally induced distortion and tilt variation at the surface would allow us to address the thickness of Titan's outer ice shell, thereby confirming the possible existence of the liquid water ocean beneath. Definitive detection of this ocean of water and ammonia under an icy layer could be provided by the Radio Science Subsystem aboard Cassini, by measuring the principal components of Titan's and Enceladus' gravitational potential. This will also provide important constraints on the satellites' internal differentiation.

If a liquid body exists in Titan's interior it could be an efficient medium for converting simple organics to complex molecules, and chondritic organic material – brought in by carbonaceous chondrites and micrometeoritic impacts – into prebiotic compounds. In addition to this kind of exogenous input which could be stored in the lakes, one should consider endogenous supply by organic syntheses occurring in the bottom of Titan's primordial oceans through the presence of hydrothermal vents like the black smokers that we discussed in Subsection 2.2.2.

But the presence of a subsurface liquid water ocean on Titan does not by itself guarantee the presence of life therein. Although at some point early in Titan's history this hypothetical ocean may have been in direct contact with the atmosphere on the one hand, and with the internal bedrock at the bottom on the other hand, presenting important analogies with the primitive Earth, the data we currently have do not tell us whether there are indeed such contacts. This kind of information would also have an impact on our understanding of the mystery of methane replenishment on Titan. A liquid water ocean containing ammonia could become buoyant and cause outgassing of methane through the crust. Such an ocean could also serve as a deep reservoir for storing methane and might possibly shelter organisms which would have to survive high pressures and concentrations of ammonia, as well as low temperatures – extreme conditions, but not dissimilar to some found on Earth. Methanogens, discussed above, have been shown to be able to exist in high concentrations of ammonia at neutral pH.

A significant geophysical difference then becomes evident when one compares Titan and Europa: on Titan, as we have seen, the liquid water layer, if real, would in all probability not currently be in contact with a silicate core. The surface of Titan appears (like Mars or Europa) an unlikely location for life at present, at least for terrestrial-type life. Nevertheless, some scientists note that Titan's internal water ocean might support terrestrial-type life, which could have been introduced there or formed early in Titan's history when liquid water was in contact with silicates. In addition, there exist photochemically derived sources of free energy on Titan's surface, which could support some lifeforms using liquid hydrocarbons as solvents. Conversely, some studies have shown that terrestrial bacteria might satisfy their energy and carbon needs by 'eating' tholin, which is abundant in Titan's atmosphere and therefore provides a means to capture the free energy from ultraviolet light and make it available for metabolic reactions.

### 4.3.1.4 Titan and the primitive Earth

There are obvious analogies between the organic chemistry activity currently occurring on Titan and the prebiotic chemistry which was once active on the primitive Earth, prior to the emergence of life.

As we saw in Chapter 2, theories of the primitive Earth suggest an oxygen-less atmosphere before the appearance of life. For example, in 2007 Laura Schaefer and Bruce Fegley predicted that Earth's early atmosphere contained $CH_4$, $H_2$, $H_2O$, $N_2$ and $NH_3$, similar to the components used in the Miller–Urey synthesis of organic compounds, often noted to be similar to the atmospheric inventory of Titan and Enceladus. However, most of the arguments in favour of the presence of a reducing greenhouse gase like ammonia in the early Earth atmosphere have now been put into perspective, as discussed extensively by Feulner et al. (2012), as it would have been destroyed by photodissociation quite quickly, and other means exist for the production of prebiotic molecules (meteoritic impacts, deep-sea hydrothermal vents, etc.). As a consequence, ammonia is no longer the favourite dominant component in the primitive Earth's atmosphere; instead, $N_2$ is now favoured, outgassing rapidly on the early Earth from the interior, creating a secondary atmosphere in which the nitrogen concentration would have quickly attained today's value. Furthermore, recent studies have looked at the heating and haze forming processes on our planet at those primitive times and found many similarities with Titan for what could have served as a primary source of organic material to the surface. Indeed, the presence of methane (at abundances as high as 100–1000 parts per million per volume) and of $CO_2$ as warming agents for the Archaean period is now advocated: this could have contributed first to warming the atmosphere in order to survive the faint young Sun and then to forming the organic haze, thus igniting an anti-greenhouse effect which would have subsequently decreased the temperature of the atmosphere and created a habitable environment on our planet. Titan offers similar possibilities in that its atmosphere is essentially composed of nitrogen of possibly the same origin and includes several thick methane haze layers, leading to the formation of an equivalent anti-greenhouse effect, albeit not caused by $CO_2$. Titan's current atmosphere is then even more suited for studying conditions on the primitive Earth.

Although Titan lacks oxygen and sufficiently elevated temperatures, unlike the primitive Earth, different evolutionary pathways on Titan may still have led to the creation of polyphenyls (these ether phenyl polymers, or complex hydrocarbons, afford excellent thermo-oxidative stability and radiation resistance). The abundances of liquid hydrocarbons on Titan are hundreds of times higher on Titan than all the oil and natural reserves on Earth. And in the atmosphere we find hydrogen cyanide and other prebiotic molecules which are among the starting materials for biosynthesis. The existence of hydrocarbons, and in particular acetylene and benzene, has really enlarged the borders of photochemical organic products. Moreover, the temporal variations that the hydrocarbon trace gases on Titan experience during a full Titan year are probably also influenced by local or regional sources and sinks.

### 4.3.2   Enceladus: water pockets far from the Sun

Saturn's geyser-spewing moon, Enceladus, is a very small satellite, only 500 km in diameter (Figure 4.15). How something so small, buried inside an ice crust, derives the energy to eject a plume 900 km out of its south pole into space is still something that puzzles scientists, who are trying to determine the heat sources that prevent this tiny moon from being frozen all over, like the others orbiting Saturn. At the same time, Enceladus poses a major challenge to traditional models of the habitable zone, since it seems to show that liquid water exists a long way (10 AU!) from the Sun, albeit underneath the surface.

Whether it is possible for life to exist in pockets or liquid water oceans underneath its surface is an even more compelling question that has interested planetary researchers since 2005, when Cassini's magnetometer first spotted the influence of the plume on the field lines before it was optically recorded by the mission's cameras. During an early flyby, Cassini passed within 97 km of the moon's surface, sampling the ejecta.

Nobody knows exactly how these plumes of gas located at or close to the 'tiger stripe' ridges near the moon's south pole are formed. Several theories have been proposed, one of them based, as for other satellites, on the gravitational pull of Saturn on this small moon,

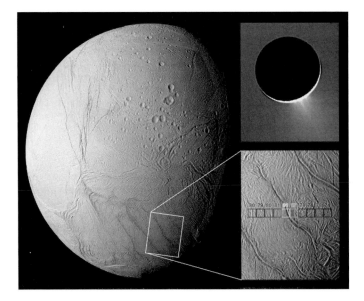

FIGURE 4.15  Surface of Enceladus, with focus on tiger stripes and an optical image of the jets emerging therefrom. The right lower zoom indicates the temperature scale across a south polar region, with the warmest temperatures detected at the 'tiger stripe' location (in yellow). For colour version, see plates section. (Image courtesy of NASA/JPL/Space Science Institute.)

perhaps causing the formation of rifts through the interior to the surface. In homage to the other geysers we know of in the Solar System, in our own Yellowstone, such models bear the names of 'Old Faithful', 'Cold Faithful' and 'Frigid Faithful' geysers. Of the two most favoured theories, one basically advocates the presence of liquid water under the surface directly supplying the plumes (Figure 4.16), while the other favours ice friction under tidal effects for their origin. The consensus tends to be that the observed heat signatures indicate that liquid water exists down there, but although this would have significant implications for astrobiology, the question is far from settled.

As an example, the 'Old Faithful' radiolytic model proposed by John Cooper and colleagues in 2009 suggests that the surface of Enceladus would be irradiated by Saturn's inner magnetospheric

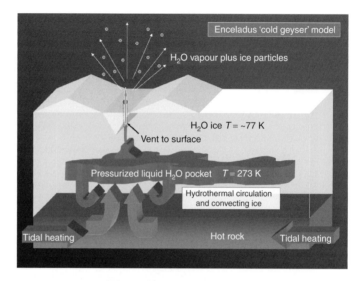

FIGURE 4.16 Enceladus 'cold geyser' interior model for the creation of the plumes observed emerging from the south pole. (Image courtesy of NASA/JPL/Space Science Institute.)

electrons, producing radiolytic oxidants. During episodic overturn of the moon's south polar terrain, these oxidants might find their way to the putative undersurface ocean where they would form carbon dioxide and other gases over long timescales, leading in good time, through some venting cracks, to cryovolcanism responsible for the plumes. This process is a means for the satellite to remain in one piece. Such a radiolytic process could also be an explanation for the geysers on Neptune's moon Triton (see hereafter), and on any icy body immersed in a radiation environment, including Eris and Haumea located beyond Pluto. Finally, the radiolytic model for Enceladus has direct implications for the habitability potential of this and other active moons like Jupiter's Europa or Ganymede, where one finds the irradiated environment but no plumes.

### 4.3.3   Future exploration of Kronian satellites

In summary, all of the satellites we have discussed in the above sections have habitable potential. Europa, Ganymede, Enceladus and

Titan may all have subsurface oceans, but we do not know how deep we need to search to get to them and whether they could really host life.

Ganymede and Europa in the Jovian system are currently major targets for finding an internal liquid water ocean among the giant planets satellites. More dedicated exploration is required before this can be established and characterized, but the astrobiological potential has been recognized for quite some time now and begs for further investigation, which is currently under way, as we have seen and discussed in Subsection 4.2.3.

In the Kronian system, Titan certainly is of considerable interest in terms of habitability and astrobiology not least because of its methanological cycle, an analogue to the terrestrial hydrological cycle, and its complex organic chemistry both in the atmosphere and on the surface. These, along with studies of Titan's interior and search for the methane reservoir, make Titan a high-priority target for future exploration if we are to understand how organic-rich worlds evolve.

In addition, the geologically active moon Enceladus and its fascinating geological features, like the south pole plumes probably caused by heated subsurface water pockets, require further long-term exploration of the Saturnian system with suitable mission components and instrumentation. Open issues have been identified for the purpose of future exploration within the Titan and Enceladus Mission (TandEM) concept (proposed to ESA as a large mission within the framework of the Cosmic Vision 2015–2025 programme, and which then became the Titan Saturn System Mission (TSSM), a joint study by the two space agencies, ESA and NASA; Coustenis *et al.*, 2009). These issues include:

– determining the organic chemistry of the two Saturnian satellites Titan and Enceladus;
– determining their present-day structure and precisely identifying whatever levels of activity they may have;
– determining whether the satellites have been subject to significant tidal deformation, and whether they possess cryovolcanism or any eruptive and seismic processes;

- identifying heat sources and internal reservoirs of volatiles (in particular methane and ammonia);
- searching for the presence of intrinsic or induced magnetic fields;
- searching for prebiotic compounds formed on Titan's surface or near subsurface. Long-term chemical evolution is impossible to study in the laboratory: *in situ* measurement of Titan's surface thus offers a unique opportunity to study some of the many processes which could have been involved in prebiotic chemistry, including isotopic and enantiomeric fractionation.

Although the Cassini–Huygens mission is a remarkable success, answering many outstanding questions about the Saturnian system and Titan in particular, it has also perhaps raised more questions. It is clear that Titan's organic chemistry and the presence of a subsurface ocean remain to be investigated. In particular, joint measurements of large-scale and mesoscale topography and gravitational field anomalies on Titan from both an orbiter and an aerial platform would impose important constraints on the thickness of the lithosphere, the presence of mass anomalies at depth and any lateral variation of the ice mantle thickness. It is astrobiologically essential to confirm the presence of such an internal ocean, even though the water layer may not be in contact with the silicate core as in Europa. However, the detection of potential biological activity in the putative liquid mantle seems challenging.

An important limitation of the Cassini–Huygens mission, as far as concerns Titan, is the insufficient spatial and temporal coverage allowed by its limited orbit around Saturn. One needs to remember that Cassini is not an orbiter dedicated to Titan or to any of the moons. For a body as special and complex as Titan, the minimum possible flyby altitude of 950 km and the uneven latitudinal coverage have impeded our attempts to explore the full set of atmospheric chemical processes. So far, opportunities for occultation have been rare, so we have not properly explored the magnetospheric downstream region. Furthermore, in spite of the wonderful opportunity offered by Huygens, we have only obtained one vertical profile of the atmosphere, and thus our understanding of horizontal transport and latitudinal variations is incomplete.

The surface of Titan, as revealed by both the Huygens probe and the Cassini orbiter, offers us an opportunity to stretch our current models in an effort to explain the presence of dunes, rivers, lakes, cryovolcanoes, ridges and mountains in a world where the rocks are composed of water-ice rather than silicates and the liquid is methane or ethane rather than liquid water, but the limited high-resolution spatial coverage restrains our view of the range of detailed geological processes on this body. The exciting results from the Huygens post-landing measurements, although providing a valuable 'ground truth', are limited to a fixed site and short timescales, and do not allow for direct subsurface access and sampling.

Several concepts for future missions that could answer such questions and more have been considered by ESA and NASA. The Titan Saturn System Mission (TSSM) (http://www.lesia.obspm.fr/cosmi cvision/tssm/tssm-public/) was considered in 2008–2009: its focus was on enhancing our understanding of Titan's and Enceladus' atmospheres, surfaces and interiors, determining the pre- and protobiotic chemistry that may be occurring on both objects, and deriving constraints on the satellites' origin and evolution, both individually and in the context of the complex Saturnian system as a whole. The mission was an ambitious and challenging combination of three elements for remote observations (by a dedicated orbiter) and *in situ* observations (with a montgolfière and a lake lander; Figure 4.17). Since then, more focused and simpler mission concepts for an orbiter, an aerial aerostat and a lake lander separately have been proposed to the space agencies, but at the time of writing, nothing has been definitively decided in terms of follow-up investigations after Cassini–Huygens.

## 4.4  COMETS

Comets and some asteroids (the most primitive ones) are of major interest for astrobiology, because, as mentioned above (Subsection 2.3.3), they might have fed the terrestrial atmosphere with prebiotic molecules in its early ages, especially at the time of the Late Heavy Bombardment. In addition, comets are the most water-rich bodies in the Solar System (about 80 per cent by mass); they are unaltered

FIGURE 4.17 The Titan Saturn System Mission concept with the dedicated orbiter, the lake lander and the floating hot-air balloon; artist's view. For colour version, see plates section. (Image by C. Waste, courtesy of NASA/JPL.)

remnants of the conditions and processes at work in the early Solar System, and their chemical composition shows striking similarities with interstellar matter. For all these reasons, these small bodies can provide precious information in our search for extraterrestrial habitats.

### 4.4.1 Comets: back to the origins

Because of their peculiar – sometimes spectacular – appearance in the sky, comets have been known from antiquity (Figure 4.18). Their origin was a mystery for scientists until Tycho Brahe (1546–1601) demonstrated, by measuring the parallax of a comet, that the apparition was not an atmospheric phenomenon. Their orbits were later determined by Isaac Newton (1644–1727) and Edmund Halley (1656–1742) who demonstrated that a comet reappeared periodically. He successfully predicted the 1758 return of a famous periodic comet. The apparition took place long after his death, and the comet was then given his name.

We now know that comets are small bodies, less than a few kilometres in size, which travel on very elliptic orbits between the

FIGURE 4.18 Comet Halley as shown on the Bayeux tapestry relating the conquest of England by William the Conqueror (1066). For colour version, see plates section. (Image courtesy of Wikipedia, http://en.wikipedia.org/wiki/Halley's_Comet.)

outer Solar System where they spend most of their time and the inner Solar System, with a typical period between a few years and a few tens or hundreds of years. They are mostly made of water and ices with some fraction of organics and rocks; because they have remained mostly unaltered since their origin, comets are precious witnesses to the conditions of formation and early evolution of the Solar System.

The observation of a comet is difficult. Far from the Sun, the comet is a cold object that only consists of its nucleus, too faint to be easily imaged or observed by spectroscopy from the ground. When the comet approaches the Sun, the surface of the nucleus sublimates under the effect of solar radiation; water and other gases evaporate, carrying jets of dust, and form the coma which scatters the Sun's light and becomes brighter and brighter (Figure 4.19). The observer can thus analyse the components of the coma – parent molecules, daughter molecules, radicals and ions resulting from solar ultraviolet irradiation – but the nucleus itself remains hidden behind the coma.

FIGURE 4.19 Comet P/Halley as taken on 8 March 1986 by W. Liller, Easter Island, part of the International Halley Watch (IHW) Large Scale Phenomena Network. For colour version, see plates section. (Image courtesy of NSSDC/NASA.)

### 4.4.2   Origin of comets: two distinct reservoirs

What is the origin of comets? Information is provided by the study of their orbits. Comets can be divided into two classes: the short-period ones (with periods less than a few tens of years) and the long-period ones (including the parabolic and hyperbolic ones that never return). By studying the orbits of long-period comets and taking into account perturbations due to the giant planets, the Dutch astronomer Jan Oort (1900–1992) demonstrated that all these objects were coming from a very distant shell – now called the Oort cloud – located at about 50 000 AU from the Sun. This discovery was later confirmed by Brian Marsden (1937–2010).

The Oort cloud could well contain some hundred billion comets or even more; however, its total mass is estimated to be just a few times the mass of the Earth. The comets were not formed in the Oort cloud itself, because there was probably not enough available material at such great distances from the Sun; most likely, they were formed in the vicinity of the giant planets' orbits and expelled outward as an effect of their gravitational perturbations (see Chapter 1). Occasionally, owing to some external gravitational perturbation, an object can be ejected from the Oort cloud, approach the inner Solar System and be captured on a periodic orbit as a result of planetary perturbations; such is the case for comet Halley. Comets coming from the Oort cloud are characterized by a long period, a large eccentricity and a random inclination.

The other main class of comets includes short-period objects with low inclinations and low eccentricities. These comets are believed to originate from the Kuiper Belt, at 40–100 AU from the Sun, and are thus called Kuiper Belt comets (Figure 4.20). The toroidal nature of the Kuiper Belt, different from the isotropic Oort cloud, explains the low eccentricity and low inclination of these objects. The study of their orbital history shows that they often spend part of their lives in orbit around Jupiter, hence their appellation of 'Jupiter-family comets'.

It is important to remember that Oort comets and Kuiper Belt comets originated from different reservoirs, located at different heliocentric distances, and their composition may reflect the associated different conditions at the time of the comets' formation. Because of their shorter periods, Kuiper Belt comets have experienced more

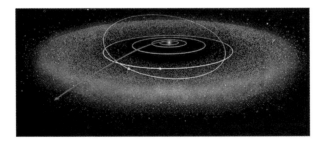

FIGURE 4.20 An artist's representation of the Kuiper Belt, beyond the orbit of Neptune. The circular orbits of Jupiter, Saturn, Uranus and Neptune, as well as the eccentric one of Pluto extending well into the Kuiper Belt, are shown for comparison. The arrow indicates the trajectory of the New Horizon mission. (Image courtesy of NASA.)

perihelion passages than the Oort cloud ones, losing material each time they approach the Sun. For this reason they are usually weaker and more difficult to observe.

In addition to these two main reservoirs, a third one has recently been proposed, in the outer part of the asteroidal main belt; these comets might have been responsible for bringing water to Earth (see Section 3.5).

### 4.4.3   What are comets made of?

Thanks to the extension of the observable spectral range accessible to astronomers, our knowledge about the composition of comets has made huge progress over the past 50 years. Until the 1950s, cometary spectroscopy was limited to the visible range, and only daughter products (mostly $C_2$, CN and CH) were known. Then, in the 1970s, ultra-violet observations revealed emissions from atoms and radicals (H, O, C, S, CS, OH...). Starting in the 1980s, infrared and millimetre observations led to the detection of parent molecules that bear key information about the chemical composition of cometary nuclei: $H_2O$, $CO_2$, CO, HCN, $H_2CO$ and others.

The return of comet Halley in 1986 provided astronomers for the first time with the opportunity to detect parent molecules. In addition to an unprecedented international ground-based observing campaign, five spacecraft encountered the comet as it crossed the ecliptic plane in March 1986 (see Subsection 2.3.3). For the first time, images of a cometary nucleus were obtained (Figure 4.21); they revealed that the nucleus was not spherical but elongated and highly irregular, partly covered with ice and partly with dark refractory material, most likely of carbonaceous origin. Water was at last unambiguously identified. Its presence as a major component had been strongly suspected, following the famous 'dirty snowball model' proposed in 1950 by the American astronomer Fred Whipple (1906–2004); the major argument in its favour was the equal abundance of H and OH, the two most abundant daughter products measured in several comets, and also the presence of the $H_2O^+$ ion.

FIGURE 4.21 The nucleus of comet Halley as imaged by the camera of Giotto in March 1986. (Image © Halley Multicolor Camera Team, Giotto Project, ESA.)

The water was identified through its infrared emission around 2.7 µm, both from space and from the Earth, using the Kuiper Airborne Observatory. Other parent molecules were detected in the infrared range: $CO_2$, CO, OCS, $H_2CO$ and more complex organic molecules, both aromatic and aliphatic; HCN was detected in the millimetre range. In addition, *in situ* measurements of the cometary gas and dust, using mass spectrometry, revealed that the cometary matter was very primitive, with large abundances of H, C, O and N. Altogether, the exploration of comet Halley demonstrated the close link between cometary and interstellar matter.

Subsequent observations of bright comets confirmed this discovery. The most spectacular event was the arrival of comet Hale–Bopp in 1997. This exceptionally bright object, coming from the Oort cloud for the first time, had a diameter of about 50 km. It was observed by ground-based telescopes, especially in the millimetre and

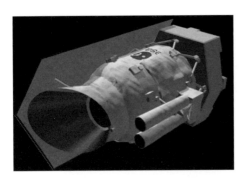

FIGURE 4.22 The Infrared Space Observatory. This infrared satellite, launched by ESA in November 1995, operated in Earth orbit until April 1997. (Image © ESA.)

submillimetre range, and by the Infrared Space Observatory (ISO; Figure 4.22) in the infrared range. ISO observations revealed the spectacular similarity in chemical composition between cometary and interstellar dust. By decreasing abundance order in cometary ices, the main parent molecules are $H_2O$ (80 per cent by number), then CO and $CO_2$ (about 10 per cent); then $CH_3OH$, $NH_3$, $CH_4$ and $H_2CO$ (about 1 per cent). About 20 gaseous parent molecules were identified in comet Hale–Bopp, all also present in the interstellar medium, including an 11-atom molecule, ethylene glycol ($HOCH_2CH_2OH$). ISO also observed the spectrum of Hale–Bopp over the whole infrared range and established the exact nature of cometary dust, dominated by forsterite, a special type of magnesium-rich olivine ($Mg_2SiO_4$), which has also been detected in the dusty envelope surrounding the star HD69830, a star which also has a system of three exoplanets. The cometary solid phase also includes carbonaceous material, as detected on comet Halley, and also, most likely, polycyclic aromatic hydrocarbons (PAHs).

Apart from the nucleus, carbonaceous grains present in the comet are a potential source for cometary molecules. Some molecules, such as CO, $H_2CO$, OCS, HCN, CN show a distributed source which could originate from the thermal degradation of grains. Laboratory simulation experiments suggest that polyoxymethylene or POM ($H_2CO)_n$, hexamethylenetramine HMT ($C_6H_{12}N_4$), HCN polymers, or carbon suboxide poymers ($C_3O_2)_n$ are plausible cometary

FIGURE 4.23 Comet Wild 2, as observed by the Stardust mission on 2 January 2004. Launched on 7 February 1999, the spacecraft collected samples of the comet's dust and returned them to Earth on 15 January 2006. (Image courtesy of NASA Discovery.)

compounds which could explain the observed distributed sources of gaseous cometary species.

Another source of information has come from the Stardust space mission. The spacecraft, launched by NASA in 1999, collected samples from the coma of the comet 81 P/Wild 2 (Figure 4.23) and returned them to Earth in 2006. Cometary and interstellar particles were collected in an aerogel substrate. Careful analyses revealed the presence of organic compounds including aliphatic chains, methylamine $CH_3NH_2$, ethylamine $CH_3CH_2NH_2$ and possibly glycine $NH_2CH_2COOH$. PAHs were also tentatively identified in Stardust samples.

In 2005, NASA performed another experiment to study the internal composition of a comet. On 4 July 2005, the Deep Impact mission sent an impactor into the surface of comet Tempel 1, creating a 30-m crater. Analyses of excavated material (which appeared to be more dust than ice) revealed clays, carbonates, sodium and crystalline silicates; the presence of clays and carbonates was unexpected as it usually implies the presence of liquid water. The spacecraft, renamed Epoxi after the Tempel 1 impact, was later retargeted toward another Kuiper Belt comet, Hartley 2, which it encountered on 4 November 2010. The comet exhibited an unexpected peanut shape, with evidence for different morphological regions, and several bright jets of water vapour and carbon dioxide, well separated in space.

It has been known for decades that the chemical composition of comets is not uniform; in particular, the dust to gas ratio varies as well as the carbon abundance, initially measured through emission lines of the $C_2$ radical in the visible range. Using the $C_2$ diagnostic, observations apparently show a carbon depletion among the Kuiper Belt comets; however, this trend does not seem to be confirmed by the abundances of their parent molecules. It is true, however, that the observation of Kuiper Belt comets in the infrared and millimetre range is much more challenging than that of Oort comets, which are typically brighter. More observations will be needed to refine this analysis.

### 4.4.4   Isotopic ratios and ortho/para ratios

Observations provide two other diagnostics about the formation conditions of comets. The first is the measurement of isotopic ratios, and in particular D/H, already mentioned earlier (Subsection 1.2.3; Figure 4.24).

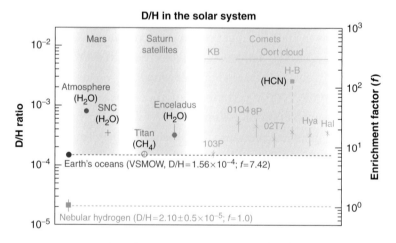

FIGURE 4.24  The D/H ratio in comets, compared with other Solar System values. KB, Kuiper Belt. (Figure adapted from Mumma and Charnley, 2011.)

We have seen that D/H in ices (and in particular in water) is enriched as a result of ion–molecule and intermolecule reactions at low temperature: at the surface of a grain, the deuterium atom, being twice as heavy as the hydrogen atom, is more easily captured. As a result, the D/H ratio is enriched in $H_2O$ ice, and the measurement of D/H in $H_2O$ in different small bodies of the Solar System provides an indication of the formation temperature of the body where it is measured (note that D/H measured in the atmospheres of the terrestrial planets tells a different story, as the D/H enrichment in Mars and Venus is due to differential escape, a totally different mechanism).

The D/H ratio was first measured in comet Halley (an Oort comet) in 1986 by mass spectrometry aboard the Giotto spacecraft, with a value of $3 \times 10^{-4}$, twice the VSMOW terrestrial value measured in the oceans. The same result was later reported on two other Oort comets, Hyakutake in 1996 and Hale–Bopp in 1997, this time using submillimetre remote sensing spectroscopy. It was then inferred that terrestrial water could not have been brought in only by comets. But the situation evolved again in 2011 with the first measurement of D/H in a Kuiper Belt comet, Hartley 2, using submillimetre spectroscopy with the Herschel spacecraft. This time D/H was found to be equal to the VSMOW value, thus reactivating the debate. Future measurements on Kuiper Belt comets are obviously necessary before a firm conclusion can be drawn (see Section 3.5).

Another precious diagnostic of cometary formation conditions is provided by the ortho/para ratio (OPR). What is this parameter? It comes from the two different possible states of the hydrogen molecule $H_2$, ortho or para, depending on the spin value $(+1/2$ or $-1/2)$ of the rotational direction of the proton of each atom. If the nuclear spins are in opposite directions, the $H_2$ molecule is called para; in the other case it is ortho. The same states apply to $H_2O$. They can be identified spectroscopically from their spectral lines, which occur at slightly different frequencies. The relative abundances of the two states can thus be measured, and they bear information on the temperature of the molecule at the time of its formation. At ambient temperature, the

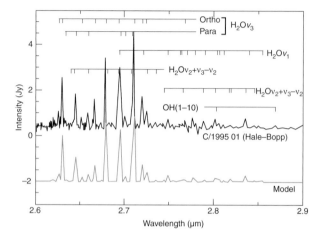

FIGURE 4.25 The spectrum of comet Halley as observed by the Infrared Space Telescope. The ortho/para ratio in cometary water is retrieved for the relative abundances of the ortho and para transitions. (Figure adapted from Crovisier *et al.*, 1997.)

ortho/para value is 3:1 (its 'normal' value), but it decreases with temperature down to very low values if the temperature is low. The OPR was first measured in comet Halley in 1986, using infrared transitions around 2.66 micrometres observed from the Kuiper Airborne Observatory (Figure 4.25). The measurement was repeated with ISO on two other comets, Hale–Bopp and Hartley 2. On these three objects (two from the Oort cloud and one from the Kuiper Belt), very low formation temperatures were measured, lower than 35 K.

Isotopic and OPR measurements are more easily made on water as it is the most abundant molecule. However, the properties of isotopic abundances and OPRs mentioned above also apply to other molecules. Nitrogen-bearing species are of special interest, because the $NH_2$ radical can be measured in the visible range. In a recent analysis by Shinnaka *et al.* (2011), the OPR in $NH_3$ has been inferred (from its ratio in $NH_2$) in 15 comets, together with the $^{15}N/^{14}N$ ratio. All objects are found with formation temperatures lower than 35 K, which confirms the water measurements.

### 4.4.5 Comets and the origin of life

The panspermia hypothesis (Box 2.2), which suggests that life might have been brought to Earth from extraterrestrial sources, has been known since antiquity. We find the first mention of it in the work of the Greek philosopher Anaxagoras five centuries BCE; a millennium later, Giordano Bruno (1548–1600) might have been influenced by this theory. More recently, Fred Hoyle (1915–2001) and Chandra Wickramasinghe (b. 1939) revived this concept. Originally, the idea was that living organisms could have travelled through interplanetary (or even interstellar) space to fertilize the Earth. However, it is unlikely that microorganisms would survive the very harsh radiation field of the interplanetary environment. A more pertinent question is whether prebiotic molecules could have been carried to Earth from outside. The answer is yes, as we know that amino acids have been found in the Murchison meteorite. Could comets have brought prebiotic molecules to Earth? Part of the answer will be obtained if prebiotic molecules are found in comets.

Scientists have placed much hope in the European Rosetta space mission (Figure 4.26), launched in 2004 and now on its way to a Kuiper Belt comet, 67-P/Churyumov–Gerasimenko. The spacecraft will orbit the comet as it approaches the Sun, monitoring the onset of its activity, and, in November 2014, will send a lander to its surface. A suite of remote sensing and *in situ* experiments will study the chemical

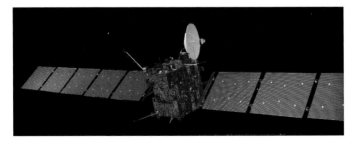

FIGURE 4.26 The Rosetta spacecraft. Launched by ESA on 2 March 2004, the spacecraft will encounter comet Churyumov–Gerasimenko early in 2014. On 10 November 2014, a lander will be deposited on the comet's surface for an *in situ* exploration of its composition and structure. (Image © ESA.)

composition of the coma and the surface, as well as the physical properties of the nucleus. This exploration is expected to help us reach a milestone in our knowledge of comets and their possible link to the appearance of life on Earth.

Could comets be habitats in themselves? Water is present in the form of ice and vapour, but, according to models of cometary interiors, the presence of liquid water is unlikely. Liquid water would require a heating source, which could be provided by the radioactivity of $^{26}$Al or possibly by the phase transition between amorphous and crystalline water; but it also requires sufficient pressure, which can be reached only inside big objects, of about 100 km in size. Kuiper Belt objects (see below) would a priori be better targets as potential habitats.

## 4.5   AT THE ORBIT OF NEPTUNE AND BEYOND

Over the past two decades, the discovery of a new class of objects beyond Neptune's orbit has opened a new window on the formation and evolution of the outer Solar System; it has also provided a link with the debris disks that are now commonly discovered around nearby stars. Major milestones were the discovery of Pluto in 1930 by Clyde Tombaugh (1906–1997), the prediction of the existence of a population of objects beyond Neptune by Kenneth Edgeworth and Gerard Kuiper in 1943 and 1950 respectively, and finally the discovery of the first Kuiper Belt object, 1992QB1, by David Jewitt and Jane Luu in 1992.

About 1350 trans-Neptunian objects or TNOs are known today. They are divided into different classes based on their orbital properties. The classical objects (about 60 per cent) have a low eccentricity and a moderate inclination; Makemake is the largest object of this category. The resonant objects (about 12 per cent) are, like Pluto, in 3:2 mean motion resonance with Neptune; the largest object of this category, after Pluto, is Orcus. The scattered objects have large eccentricities and perihelion distances close to Neptune's orbit; a typical representative is Eris whose size is very similar to (if not larger than) the size of Pluto. Finally, the detached objects have large eccentricities ($e > 0.24$) and perihelion distances beyond Neptune's orbit. The origin of this last category is still unclear; a representative object is Sedna, with a semi-major axis of about 500 AU.

### 4.5.1   Cryovolcanic Triton

Further out, around Neptune, revolves another satellite, Triton, with unique characteristics among which are an atmosphere and geysers (Figure 4.27). Triton's characteristics, in particular its large size (it is the biggest of the Neptunian moons) and its retrograde orbit, suggest that it is an object from the Kuiper Belt region captured around the planet during one of its excursions. The Voyager 2 flyby in 1989 revealed that Triton has a nitrogen and methane atmosphere, extending 950 kilometres above the surface, with a ground pressure of 14 microbars. The

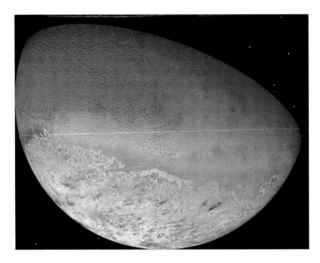

FIGURE 4.27  The surface of Neptune's largest moon, Triton, at its sub-Neptunian hemisphere. This image is a false-colour mosaic taken in 1989 by NASA's Voyager 2 spacecraft. It is one of only three objects in the Solar System known to have a nitrogen-dominated atmosphere (the others are Earth and Saturn's moon, Titan). Its frozen surface is made mainly of nitrogen ice, with some methane ice at the south pole (bright, pinkish terrain in the lower part of the image). Voyager 2 observed dust deposits left by nitrogen gas geysers (dark maculae shown as dark spots on the image). The smoother greenish region above includes the Triton 'cantaloupe terrain' and cryovolcanic and tectonic features. For colour version, see plates section. (Image courtesy of NASA/JPL/USGS.)

surface temperature is about 35.6 K. The moon's surface also has nitrogen, methane and carbon dioxide ices. As Triton's density is high, it is suspected that it has a large core of silicate rock.

The atmospheric temperature profile shows a troposphere (lower part of the atmosphere just above the surface) formed at around 8 km in altitude owing to turbulence at the surface, but no stratosphere above. Work in recent years by Candice Hansen and William McKinnon has determined the atmospheric and surface structure of Triton. A well-structured thermosphere (where heat is transported by conduction), an ionosphere and an exosphere also exist. The exospheric temperature reaches 95 K. The Hubble Space Telescope and occultation observations have shown that the atmosphere is actually even denser than suggested by Voyager 2, and that the temperature is increasing, but the mechanism for this is unknown. The lower atmosphere of Triton contains a variety of condensates. Most of the atmosphere contains a diffuse haze which probably consists of hydrocarbons and nitriles produced by the photolysis of nitrogen and methane. Discrete clouds can be distinguished at the limbs, most probably consisting of condensed nitrogen. The plumes seen by Voyager 2 near 50° S rising to altitudes as high as 8 km, and named Mahilani and Hili, could be ejections of liquid water, or could be purely atmospheric phenomena with the same composition as the rest of the atmosphere, with nitrogen rather than methane evaporating into the atmosphere and subsequently condensing (Figure 4.28).

In addition to these exciting atmospheric phenomena, it has recently been hypothesized that the capture of Triton and tidal friction during its subsequent evolution towards a circular orbit might have led to the formation of a subsurface ocean. Such a liquid ocean might have formed between the rocky core and an icy surface shell, and scientists think that it could have survived until now, but the question is how. Although radiogenic heating (heat caused by the decay of radioactive isotopes within a moon or planet, which can create heat for billions of years) contributes several times more heat to Triton's interior than tidal heating (friction from tides), radiogenic heat alone is not enough

FIGURE 4.28 Artist's view of Triton's geysers. Triton is scarred by enormous cracks. Voyager 2 images showed active geyser-like eruptions spewing nitrogen gas and dark dust particles several kilometres into the atmosphere. For colour version, see plates section. (Image © Ron Miller, courtesy of International Space Art Network, http://spaceart1.ning.com/photo/triton-geyser.)

to keep the subsurface ocean in a liquid state over long periods, and certainly not for the lifetime of the Solar System (4.5 billion years). However, tidal dissipation focuses the heat towards the bottom of the ice shell, causing the growth and expansion of the ice to slow down and even to stop. This tidal dissipation is stronger for larger values of eccentricity, meaning it would have played an even more important part in the past in maintaining the liquid water–ammonia ocean. However, several uncertainties remain in our knowledge of this body.

For instance, we do not know with certainty when Triton was captured by Neptune or how long it took for its orbit afterwards to become almost perfectly circular, as it is today. This circularization takes place thanks to tidal heating effects that we must study further to understand how they affect the internal structure. From this we derive information on a possible subsurface liquid water ocean and its depth, which could vary as the tidal forces are not constant across the globe but tend to focus on the poles in some cases. Another uncertainty is the size of Triton's putative ocean, which depends on the size of the rocky core (also not known with precision); the larger the core, the more radiogenic heating is available, thereby augmenting the size of any existing ocean. As we have seen previously in this chapter, current estimates suggest that icy bodies of the outer Solar System could contain as much as 15–20 per cent ammonia. When ammonia is mixed in a liquid, it helps to lower the temperatures at which the liquid (here the water) can remain unfrozen, and thus favours the existence of a liquid ocean underneath the surface for all these cold objects.

Recent research by Jodi Gaeman, Saswata Hier-Majumder, and James Roberts (Gaeman *et al.*, 2012) has shown that the ammonia that might be present in a putative Triton subsurface water ocean would act to lower its freezing point and maintain it in liquid form, possibly making it suitable for life. But such an ocean would probably still be very cold, about 176 K or –97 °C). As in the case of Titan and Enceladus, this would considerably slow down biochemical reactions and prevent evolution. However, as discussed in Chapter 2, enzymes working at temperatures as low as 170 K exist on Earth, and they have been shown to reverse this tendency and make the biochemical reactions more efficient.

Another (albeit more remote) possibility discussed by these scientists is that Triton might host a different type of life, one based on silicon rather than carbon (see Subsection 2.2.3). Silicon in the form of silanes is more apt to survive in low temperatures, but all this remains hypothetical at the time of writing.

Thus, in spite of many similarities and interactions, the icy satellites that we have examined in these sections are very different bodies with different exchange processes that need to be further investigated.

### 4.5.2 Trans-Neptunian objects

The exploration of trans-Neptunian objects (TNOs) is obviously limited to remote sensing analysis and spectroscopy, until the NASA space mission New Horizons, launched in 2006, encounters Pluto and its satellite Charon in 2015, and possibly another TNO later on. The surface composition of about 200 TNOs has been analysed by photometry in the visible and near-infrared range, and the brightest TNOs have also been studied by spectro-photometry at the same wavelengths. For the Pluto–Charon system we also have information from mutual occultation events and stellar occultation (Figure 4.29).

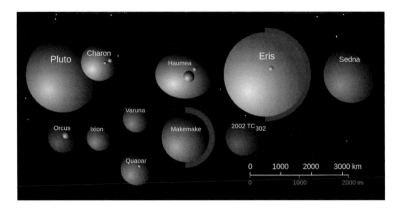

FIGURE 4.29 Some of the largest trans-Neptunian objects. The arcs around Makemake and Eris indicate the uncertainties in the size, given the unknown albedo. For colour version, see plates section. (Image courtesy of Wikimedia Commons.)

Colour photometry seems to indicate a correlation between the colour, the size and the inclination, with the smaller objects of low inclination and low eccentricity being significantly redder in most cases; this population could be more primordial than the other one. Spectroscopy of the brightest TNOs indicates a surface composition dominated by $CH_4$ and $N_2$ with, in some cases, CO, hydrocarbons ($C_2H_2$ and $C_2H_6$) and water.

What can TNOs tell us about the early history of the Solar System? Their orbital properties allow us to decipher their dynamical history, which is of key importance to understand the migration of the giant planets and also provides an explanation to the origin of the Late Heavy Bombardment. Their surface and interior properties also provide information about the physical processes that they have encountered since their origin. TNOs are considered to be pristine objects, but they have been subject to many external and internal modifying processes. First, their surface has been submitted to cosmic-ray bombardment, leading to an alteration of their molecular composition. Laboratory simulation experiments have shown that an irradiation mantle is formed, molecules in ice are broken and radicals are formed, hydrogen escapes, and a carbonaceous dark material is formed; the same layer was observed on the surface of comet Halley. Second, collisions between TNOs play an important role on the surface but also in the interior if the object is disrupted. Finally, internal processes due to short-lived radiogenic elements or accretion-generated heating may take place in the case of the largest objects. These might be responsible for the presence of crystalline water found at the surface of some objects, and not expected in view of its permanent irradiation. The water could be due to cryovolcanism, as observed on Neptune's satellite Triton; another possible origin could be micrometeoritic impacts.

What is the relevance of TNOs for astrobiology? These distant objects are too far from the Earth to have contributed to its meteoritic bombardment, as comets and asteroids did. But, as we saw for Triton, their interior probably contains a significant amount of water and, in the case of the largest objects, their internal heat (both from accretion and radioactive decay) might have been sufficient for water to be liquid. In this case the largest TNOs could possibly be considered, like some outer satellites as Triton, as potential habitats.

# 5 A revolution in astronomy: the exploration of extrasolar planets

## 5.1 FROM DREAM TO REALITY

Are there inhabited worlds elsewhere in the Universe? The question is as old as humanity. We can trace such debates back to antiquity, in texts written by Greek philosophers such as Epicurus (341–270 BCE) in particular. At the time of the Copernican revolution, a new dimension was reached, this time more on astronomical and physical grounds: since the Earth was no longer seen as the centre of the Universe, other planetary systems could exist around other stars. Giordano Bruno (1548–1600) was among the first to express his support for this new astronomical theory, in opposition to the Catholic church, a conviction for which he paid with his life. Many scientists such as Galileo (1564–1642) and Huygens (1629–1695) supported this hypothesis. Closer to our times, philosophers such as Fontenelle (1657–1757) and Kant (1724–1804), scientists such as Laplace (1749–1827) and later Flammarion (1842–1925) raised the question of the plurality of worlds.

The search for planets around other stars – also called 'extrasolar planets' or 'exoplanets' – did not start in earnest, however, until the twentieth century, because of our inability to observe them. Indeed, it is extremely difficult to detect the intrinsic visible light of such a planet, hidden in the blinding brightness of its host star, which is about ten million times brighter. Imaging extrasolar planets directly, in a few very favourable cases, has only become possible during the past decade, thanks to the development of techniques such as coronagraphy (which blocks light from the centre of a telescope in order to image the fainter surroundings) and adaptive optics (see Subsection 2.4.3). During the twentieth century, indirect methods had to be developed.

The idea is the following: the light of the exoplanet is too weak to be detectable, but the presence of the planet induces a small motion of the host star around the centre of gravity of the combined star–planet system. The first method used by astronomers to detect this motion was astrometry, the measurement of stellar positions relative to their background. It was successfully applied by Bessel (1784–1846) who first detected a low-mass companion around Sirius A, the brightest star in our skies. The companion turned out to be a white dwarf, named Sirius B. A century later, the same technique was used to search for exoplanets. In 1964, after an observing campaign of several decades, the Dutch astronomer Peter van de Kamp (1901–1995) reported the detection of a planet around the Barnard star. Ten years later, however, it was discovered that this result was an artefact caused by instrumental errors associated with the telescope. Nevertheless, Peter van de Kamp can still be seen as a pioneer in this research field.

### 5.1.1   The key to success: velocimetry

At the end of the twentieth century, astrometry was still not sensitive enough to detect exoplanets, so another technique was used: 'velocimetry', or the measurement of the relative speed of the host star as it moves around the centre of gravity of the star–planet system.

As for astrometry, the velocimetry technique was developed to search for low-mass stellar companions. Observing campaigns focused on giant planets located at quite some distance from their star (as in the structure of the Solar System) and thus needed observing periods of a decade or more to cover a complete planetary revolution.

Astronomers know how to measure very precisely the relative speed of a celestial object with respect to the observer: they use the 'Doppler effect' which slightly shifts the spectrum of the object (Box 5.1). If the object approaches the observer, the light is shifted towards the blue (higher frequencies, or shorter wavelengths); if it recedes, the shift is towards the red (lower frequencies, or longer wavelengths). To measure this effect, astronomers use high-resolution spectrometers which measure the Doppler shift of a large number of spectral

BOX 5.1   **The velocimetry (or radial velocity) technique**

The major problem in exoplanetary detection is the very high flux contrast between a star and the planet orbiting it. In the case of the Sun–Jupiter system, the flux contrast, in the optical range, is about 10 million – and the angular distance on the sky, as seen from the distance of the nearest stars (about 10 parsec) is only 0.5 arcsec! The light of the planet is thus completely hidden in that of its host star. For this reasons, astronomers of the twentieth century have searched for indirect methods: if a star has a companion such as a big planet, its motion must be disturbed accordingly, both star and planets orbiting around the centre of gravity of the system.

In principle, this motion may be detected by two methods: astrometry, which detects stellar motions in the sky plane, and velocimetry, which detects changes of velocity relative to the observer (this is called the radial velocity). Astrometry was used first but was unsuccessful, because it required an accuracy in the stellar motion measurements which could not be achieved at that time. In contrast, velocity measurement, already successfully used for detecting low-mass stellar companions, demonstrated in the 1990s its ability to detect giant exoplanets orbiting their host star.

How can we measure the radial velocity of a celestial body? Astronomers use a well-known property associated with the speed of light: if an object is moving relative to the observer, the light emitted by the object (at any wavelength or frequency, from the ultraviolet to the radio range) is shifted toward the blue (or higher frequencies) if the object approaches the observer, and toward the red (or lower frequencies) in the opposite direction. This very important effect was used by astronomers to determine that galaxies are receding, thus showing evidence for the expansion of the Universe. The same property can be used to measure very accurately the radial velocity of stars. The visible spectral range is used, because it contains a large number of spectral lines (due to the atmospheric constituents present in the star), all exhibiting Doppler shifts which can be measured and combined. High-resolution spectrographs are used and measurements are systematically performed as a function of

BOX 5.1    **(cont.)**

time, in order to determine any variation in the radial velocity of the star. If a planet is present on a circular orbit around the star, this curve shows a sinusoidal modulation associated with the revolution period of the planet. In the case of the Sun–Jupiter system, the presence of Jupiter induces a modulation of 12.5 m s$^{-1}$, within the detection capabilities of the high-resolution spectrographs operating in the 1990s.

If we were to detect the presence of a Jupiter-like exoplanet around a nearby star, we would have to wait for 12 years for it to complete a full revolution. This would require a long-term monitoring programme, and such campaigns were actually set up for this purpose. But exoplanet discoverers have benefited from an exceptionally favourable circumstance that was not at all anticipated: many large giant planets orbit very close to their host star, with a revolution period of only a few days. The first exoplanet detected around a solar-type star was identified by velocimetry: the discovery of 51 Peg b was announced by Michel Mayor and Didier Queloz, from the Observatoire de Genève, in August 1995. With a mass of 0.45 Jovian masses, the planet orbits at 0.05 AU from its star, with a revolution period of 4.2 days. Observations were taken with the ELODIE high-resolution spectrometer at the 1.93-m telescope of Haute-Provence Observatory. In the following weeks, two other exoplanets were discovered by the US astronomers Geoffrey Marcy and Paul Butler: 47 UMa b and 70 Vir b, with masses of 2.6 and 7 times that of Jupiter, which are located at 0.5 and 2 AU away from their host stars, respectively. Other discoveries were reported over the next few months, showing evidence for a large population of giant exoplanets orbiting very close to their stars with periods of a few days.

The velocimetry method has thus been a pioneering means of discovering exoplanets. Still, it does have some limitations. The first is due to uncertainty about the inclination of the system as seen from the observer. If the orbital plane includes the line of sight between the star and the observer (the star–observer axis), the modulations of the radial velocity are maximum; in contrast, if this line is perpendicular to the orbital plane, no modulation is observed (this case is best suited for astrometry studies). In an intermediate case, the amplitude $A$ of the observed modulation of the radial velocity is given by the following formula:

BOX 5.1  **(cont.)**

$$A = 28.4 \, P^{-1/3} \, (M_P \sin i) \, M_S^{-2/3}$$

where $P$ is the revolution period of the planet (in years), $M_P$ is the mass in Jovian masses, and $M_S$ is the mass of the star in solar masses. It can be seen that only a lower limit of the planet's mass is inferred.

There is an obvious observational bias associated with the velocimetry method. The detection of giant planets close to their stars is definitely favoured; indeed, these objects were among the first ones to be discovered. Since the first discovery of 51 Peg b, huge progress has been

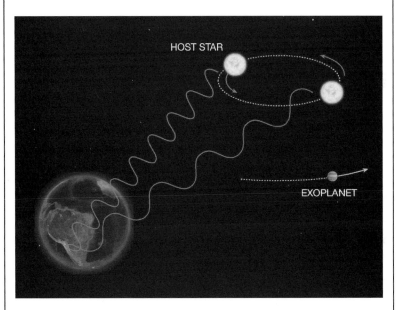

FIGURE BOX 5.1  The velocimetry method to detect exoplanets. The radial velocity of the star is measured as it is modulated by the star's motion with respect to the centre of gravity of the star-planet system. As the star moves toward the Earth, the signal emitted by the star is blue-shifted, i.e. moved towards higher frequencies (or shorter wavelengths); as it moves back in the other direction, the signal is red-shifted, i.e. moved towards lower frequencies (longer wavelengths). The velocity method allows astronomers nowadays to detect radial velocities lower than 1 m s$^{-1}$, bringing the domain of Earth-like planets within detectability limits. For colour version, see plates section. (Image © ESO.)

BOX 5.1   **(cont.)**

achieved in the development of specific high-resolution spectrometers devoted to this research. The HARPS spectrometer, operating on the 3.6-m telescope of the European Southern Observatory (ESO) in La Silla, has been leading the field. Detection limits in terms of velocimetry measurements have decreased from about 10 m s$^{-1}$ in the 1990s to about 1 m s$^{-1}$ now. As a result, the velocimetry technique is now able to reach the domain of Earth-like planets – an old dream for all astronomers.

transitions (typical of the various elements present in stellar atmospheres). By measuring systematically the Doppler effect of a given star over a very long period of time (several years), astronomers can hope to detect a regular, periodic motion, diagnostic of the presence of a low-mass stellar or planetary companion.

Starting in the 1990s, several teams (in particular Geoffrey Marcy and Paul Butler in the USA, Bruce Campbell and Gordon Walker in Canada, and Michel Mayor and Didier Queloz in Switzerland) made a systematic search for exoplanets around nearby solar-type stars. Why solar-type stars? Because the ultimate quest, beyond the discovery of an exoplanet, is the search for a habitable world. In the absence of precise information about possible forms of life different from our own, astronomers prefer to search for objects more or less similar to Solar System planets. In August 1995 came the explosive news of the discovery of the first exoplanet around a solar-type star by Michel Mayor and Didier Queloz (1995) (Figure 5.1, 5.2). Using a high-resolution spectrometer at the 1.93-m telescope of Haute-Provence Observatory (Figure 5.3), they discovered a Jupiter-like exoplanet (its mass is at least half the Jovian mass), orbiting about a solar-type star, 51 Peg. It had an astonishingly short orbital period of only 4 days (Figure 5.2), and an orbital distance of just 0.05 AU.

(a)

(b)

FIGURE 5.1 The first discoverers of exoplanets. (a) The Polish astronomer Alexander Wolszczan announced in 1992 the discovery of two small planets around a millisecond pulsar, PSR 1257 + 12. Image credit: C. Czaplinski. (b) Michel Mayor (right) and Didier Queloz (left), the discoverers of 51 Peg, the first exoplanet around a solar-type star, in 1995, at ESO's 3.6-metre telescope at La Silla Observatory in Chile. The telescope hosts HARPS, used for exoplanetary discoveries. (Image courtesy of ESO, © L. Weinstein/Ciel et Espace Photos.)

It should be noted up front that 51 Peg b, as the object is now known, is not the first exoplanet detected by astronomers. In 1992, using a completely different technique, the Polish astronomer Alexander Wolszczan (Figure 5.1a) announced the detection of two

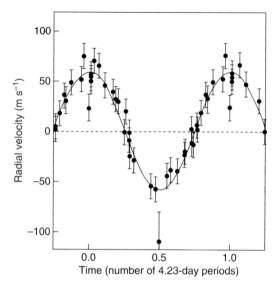

FIGURE 5.2 Variations of the radial velocity of 51 Peg, showing the periodicity associated to the presence of a planet, 51 Peg b. (Redrawn from Mayor and Queloz, 1995.)

small planets around a pulsar, PSR 1257 + 12. Pulsars, or 'pulsating stars', are neutron stars which represent the ultimate stage of stellar evolution, after the supergiant and supernova phases. These extremely dense objects are rotating at high speed (around once per millisecond in the present case) and emit a very strong radio signal, modulated by the pulsar rotation. The presence of planets around a pulsar induces a slight change in the periodicity of this radio signal; such an effect was detected on PSR 1257 +12. How could planets form around a pulsar? Probably within a disk when the stellar matter has been accreted by the pulsar. Such objects are likely to have little in common with the planets as we know them, and the impact of this spectacular astronomical result is quite limited for astrobiology.

## 5.1.2   Giant exoplanets close to their stars

The impact of the discovery of 51 Peg b was extraordinary. First, it demonstrated that planets can exist around other solar-type stars.

FIGURE 5.3 The 193-cm telescope of Haute Provence Observatory where 51 Peg b was discovered. (Image courtesy of Haute Provence Observatory, via Wikimedia Commons.)

Second, the configuration of this planetary system was extremely surprising: try to imagine Jupiter orbiting the Sun in 4 days, at a distance of 0.05 AU (ten solar radii)! Not only was the new planetary system unusual; it also questioned what we had understood about the formation scenario of our own Solar System, where giant planets can only form at large distances from the Sun. Our perceptions had to change.

Once the path and the minds were opened, several other discoveries followed rapidly. Other detections of giant exoplanets orbiting close to their host star were reported within weeks by Geoffrey Marcy's team, who confirmed the 51 Peg b detection and announced the discovery of two planets orbiting solar-type stars, 47 UMa and 70 Vir. With masses of at least 2.6 and 7 times the Jovian mass, the planets

FIGURE 5.4 The HARPS instrument at ESO (La Silla), used for radial velocity measurements. The spectrometer operates under vacuum to ensure maximum stability of the measurements. (Image © ESO.)

are located at about 0.5 and 2 AU from their host star. Other results soon confirmed this peculiar behaviour. Thus, by the end of 1995, two major results were established: first, the Solar System is not unique; and second, almost as importantly, its formation and evolution scenario cannot be systematically applied to other planetary systems (Figure 5.4).

### 5.1.3   Formation and migration in planetary systems

This odd configuration of many planetary systems where giant planets are located in the immediate vicinity of their stars came as a surprise to many a scientist. Since the early work of Kant and Laplace in the nineteenth century, it had been suggested that the formation of the Solar System took place within a 'primordial nebula' which collapsed into a disk. Planets accreted within this disk from solid particles, by coalescence, then by mutual collisions, and finally by gravity. After about 10 million years, the disk (with all the gas and the smallest particles) dissipated during an intense activity phase (called the 'T-Tauri phase') of the young star (see Section 1.2).

We have seen in Chapter 1 how the nucleation model naturally leads to the formation of two classes of planets, the terrestrial ones and the giant ones. Near the Sun, only the heaviest materials (mostly silicates, metals and metallic oxides) were in solid form. Because they are made of heavy atoms, according to the general law of cosmic abundances, such materials were relatively rare. Rocky planets – the terrestrial planets – formed by this process were consequently small and dense. In contrast, at larger heliocentric distances (at temperatures below about 200 K), the environment was cold enough for light molecules ($H_2O$, $CH_4$, $NH_3$, $CO_2$...) to be in the form of ices. These molecules, much more abundant than the heavier elements, were incorporated into icy cores which became big enough to capture by gravity the surrounding gas, mostly composed of hydrogen and helium. This process led to the formation of giant planets, with a large mass, a very large volume and a low density. Satellites were formed in the equatorial plane of these planets, within the disk formed after the collapse phase of the surrounding nebula; they are the so-called 'regular' satellites of the giant planets.

Over the past few decades, astronomical observations from the ground (using infrared and millimetre telescopes) and from space (in particular the Hubble Space Observatory) have shown that a large number, probably more than 50 per cent, of very young stars (less than 10 Myr old) are surrounded by a disk (Figure 5.5). Star formation through the collapse of a molecular rotating cloud into a disk thus appears to be a common scenario, and planetary formation within the disk is expected to take place via the same processes as in the Solar System. Assuming cosmic abundances in all cases, one might thus expect to find giant exoplanets far from their host stars.

How can we then explain the presence of giant exoplanets near their host star? Very early after the first discoveries, a mechanism was suggested: migration. In this scenario, giant exoplanets are formed at greater distances from their star, as in the Solar System, but they migrate inward toward the star because of an interaction between the planet and the disk (see Section 1.2). But some questions remain unanswered. What is the mechanism that stops the migration at a distance of about 0.04 AU from the star, where many giant exoplanets are found, and prevents them from crashing into the star? Possibly there is an inner edge to the protoplanetary disk at this distance, and the interaction between planet and disk stops because of the lack of matter. Another question concerns the Solar System: was there any inward migration of the giant planets in this case?

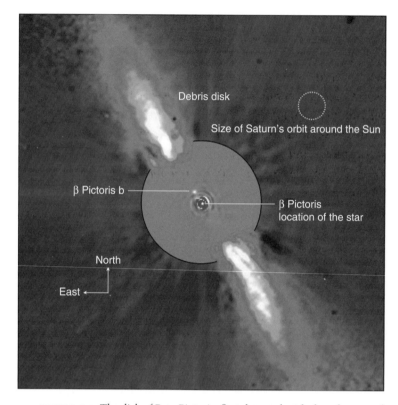

FIGURE 5.5   The disk of Beta Pictoris, first detected with the telescope of Las Campanas. This was the first identification of a debris disk around a young star. In 2009, a planet was discovered close to the star, orbiting in the plane of the debris disk. For colour version, see plates section.
(Image © ESO/A.-M. Lagrange *et al.*) (Smith & Terrile, 1984)

Recent numerical simulations, performed in particular at the Nice Observatory, have shown that even in our case, some migration of the giant planets is most likely. Jupiter and Saturn may have been closer to each other in the past. The planets separated (Jupiter moving inward and Saturn outward) while the Jupiter–Saturn system moved toward the 2:1 resonance (with Jupiter's revolution period being half that of Saturn). Uranus and Neptune were probably formed closer to the Sun than they are now, at about 15–20 AU, so that they could acquire enough material to form their icy core; Neptune, bigger than Uranus, might have been initially closer to the Sun than Uranus (see Section 1.2).

In summary, the migration scenario within the Solar System is not fully demonstrated, but numerical simulations show that it is a possible dynamical evolution of the system. It appears an interesting possibility which would explain many of the dynamical features of several families of Solar System objects, including asteroids and the Kuiper Belt. Similarly, it suggests that migration might be a general mechanism at work in protoplanetary disks, with different issues according to the various systems. It also illustrates how our knowledge of the Solar System history can benefit from a comparative study between other planetary systems and ours.

### 5.1.4 How to detect exoplanets from planetary transits

Four years after the detection of 51 Peg b, a major new result was achieved: the first detection of an exoplanet around a solar-type star, by David Charbonneau and his colleagues (2000), using the transit method and ground-based stellar photometry. What is this all about? The transit method is another indirect way of detecting exoplanets. It consists in measuring the light of the star when, by chance, a planet crosses the field of view in front of the star. At that point, the stellar flux is slightly decreased because of the planetary occultation. If the Sun were observed from outside the Solar System, its occultation by Jupiter (a tenth the diameter of the Sun) would induce a decrease of the solar flux by 1 per cent; occultation by the Earth (about 100th the diameter of the Sun) would induce a decrease by 0.01 per cent. By measuring the stellar light very precisely, before, during and after a planetary transit, and by repeating the observation over a large number of transits, it is possible to detect the presence of the planet and, if we know the stellar diameter, to derive the planetary diameter (Box 5.2).

Of course, this technique requires a special geometry: the Earth must be located in the plane of the planet's orbit or close to it. This probability is higher if the exoplanet is big and close to its star. Fortunately, this is exactly the case for the first exoplanets discovered

BOX 5.2   **The method of planetary transits**

Transits of terrestrial planets in front of the Sun have been known for centuries. Transits of Venus across the face of the Sun have been used since the eighteenth century to determine the distance of Venus, and hence the Sun–Earth distance and all distances in the Solar System. The same method has been successfully used to detect exoplanets around nearby stars. The only condition to be respected is that the Earth must be close to the orbital plane of the exoplanet. This condition is more easily fulfilled when exoplanets are located close to their host star – often the case for exoplanets, as we have seen before.

When an exoplanet passes in front of its host star, the star is occulted on a small fraction of its disk, and the stellar flux is decreased accordingly. For a Jupiter-like planet transiting the Sun, the solar flux is decreased by 1 per cent, as the radius of Jupiter is about a tenth of the solar radius. In the case of the Earth, the decreasing factor would be only 0.01 per cent.

The method of transits consists in a continuous photometric monitoring of a star field over a long period of time – at least weeks, or, even better, months. In order to detect giant exoplanets, an accuracy of $10^{-3}$ is required, which is currently achievable for high-precision ground-based photometry. Giant exoplanets can thus be detected with this technique: this was the case for HD 209458b, the first exoplanet to be identified during a ground-based transit (Charbonneau *et al.*, 2000). The transit was later observed with higher precision using the Hubble Space Telescope.

Planetary transits offer more than the 'primary' transit, when the planet passes in front of its host star. This first measurement provides us with a determination of the planetary radius and, coupled with velocimetry measurements, a determination of its inclination, mass and density. Both methods are thus highly complementary, and their combination has led to an enormous increase of our understanding of exoplanets. Actually, following the identification of potential exoplanet candidates after primary transit measurements, velocimetry measurements are required in all cases to confirm the existence of the exoplanet and remove the possibility of 'false positives' due, in particular, to stellar variability. Another event occurs when the planet passes behind

BOX 5.2  **(cont.)**

the star: the event is called a 'secondary transit' or, more correctly, an occultation. In this case, a direct detection of the exoplanet can be achieved, the planetary emission being the difference between the flux (from star + planet) just before or just after the transit, and the flux of the star alone, during the secondary transit.

Both the primary and the secondary transits are important as they lead to complementary information. The primary transit allows us to probe the exoplanet's atmosphere at the terminator (the dividing line between the day side and the night side). The flux decrease is especially strong if the exoplanet has a hot, extended atmosphere. In contrast, the intrinsic emission of the exoplanet is measured by difference at the time of secondary transits. The radiation can be either stellar reflected flux (at shorter wavelengths) or intrinsic thermal emission. In the case of a hot Jupiter observed during secondary transits, thermal radiation dominates in the infrared range beyond about 1 μm.

The planetary transit technique has been boosted by the launch of two space missions: first CoRoT, in 2006, and later Kepler, launched in 2009. Kepler has the sensitivity for finding exoplanets around weak stars, many of them too weak to be followed by velocimetry. Several tens of exoplanets have been confirmed, and thousands of Kepler candidates have been identified.

The observation of exoplanets by transit has opened a new field of research: spectroscopy of exoplanets' atmospheres at the time of transit. During primary transits, atmospheric compounds can be detected in transmission, which gives access to their column density along the line of sight. During secondary transits, the true emission of the day side is observed. Reflected starlight may dominate at short wavelengths, again giving access to the identification of minor species and an estimate of their column densities. At longer wavelengths, when thermal emission dominates, the spectral analysis is more complex, as it depends upon the thermal vertical structure of the exosphere which is not known a priori. A careful analysis must be performed to disentangle the thermal profile from the vertical distributions of the atmospheric constituents.

So far, transit spectroscopy of exoplanets' atmospheres has been mostly achieved on the two brightest targets, HD 209458b and HD

BOX 5.2 **(cont.)**

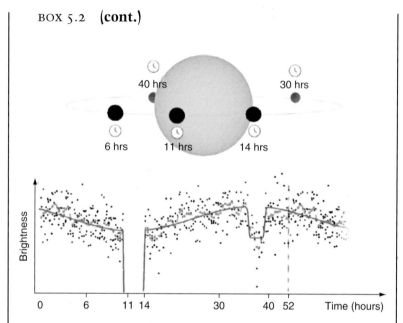

FIGURE BOX 5.2 The two types of planetary transits: (1) the primary transit, when the planet passes in front of its host star; (2) the secondary transit, when the planet passes behind it. The secondary transit induces a decrease of the stellar flux which is typically weaker than the primary transit. Just before and after the secondary transit, the total flux is higher due to the contribution of the planet's emission. (Figure adapted from Tinetti *et al.*, 2012.)

189733b. Several atoms (H, O, Na, K, Cr. . .) have been detected by primary transit in the ultraviolet and visible range, and several molecules ($H_2O$, $CH_4$, CO, $CO_2$. . .) have been found using both primary and secondary transits.

by velocimetry. The transit method is thus especially well suited to this study. In addition, detecting flux variations of the order of 1 per cent is within the capabilities of ground-based stellar photometry. This explains the first successful detection by Charbonneau in 1999; soon after, the transit curve of this exoplanet – now famous under the unfriendly name of HD 209458 – was confirmed with much higher precision using the Hubble Space Telescope (Figure 5.6).

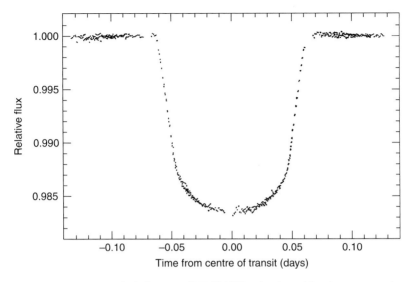

FIGURE 5.6 The lightcurve of HD 209458 at the time of the planetary transit. The image was taken from the HST. (Redrawn from Brown *et al.*, 2001.)

At the beginning of the 21st century, most exoplanets had been detected by velocimetry, and most of them were giant objects. Scientists even thought at that time that the powerful radial velocity method would not have the capability to reach low-mass exoplanets, those of about 10 terrestrial masses, expected to be too small to attract the surrounding nebula. Indeed, it was considered that the stellar variability or the lack of stability of the spectrometer would limit the sensitivity of the method to planets above this threshold. Actually these predictions appear to have been too pessimistic: at the end of the 2000s, small exoplanets of only a few terrestrial masses have been discovered by velocimetry. The search for such exoplanets (also called 'super-Earths') developed very actively during this decade. Ground-based networks were set up to search for giant exoplanets, and space missions were designed for super-Earths and even exo-Earths, with photometric sensitivity better than 0.01 per cent.

The first space mission designed for this purpose was the CoRoT mission, developed and launched by the French CNES agency, in operation in Earth orbit until November 2012 (Figure 5.7). Initially delayed because of budget problems, CoRoT was finally launched

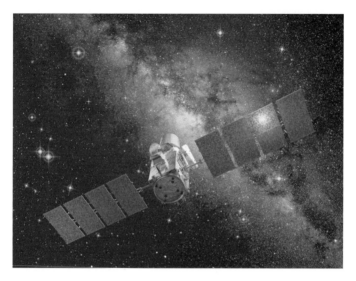

FIGURE 5.7   The CoRoT space mission. CoRoT was launched by CNES into Earth orbit in December 2006 with the objective of detecting exoplanets in transit. After 6 years of operation, CoRoT has detected about 25 exoplanets, and many more are waiting confirmation. For colour version, see plates section. (Image © CNES/D. Ducros.)

on 27 December 2006 with a double objective: to study stellar oscillations and search for exoplanets by transit. It so happens that these two very distinct objectives require a similar technique: the photometric survey of a star field over a long time (several months) with a photometric sensitivity better than 0.01 per cent. After several years in operation, the CoRoT mission has proven a great success for both scientific programmes. Several tens of candidate exoplanets – and possibly more – have been identified. However, their definite confirmation requires their identification by velocimetry, which implies a lot of effort and telescope time. Ground-based programmes have thus been set up as a follow up of the CoRoT observations. Currently, more than 25 exoplanets have been definitely identified by CoRoT, and the list will no doubt be extended in the coming years as several more candidates await confirmation.

After CoRoT came Kepler, a more ambitious and extremely successful NASA mission, launched in March 2009 and operating in

FIGURE 5.8 The Kepler space mission. Launched in March 2009, the Kepler satellite is designed for detecting exoplanets by transit. Kepler has detected about 100 confirmed exoplanets and more than 2000 possible candidates. For colour version, see plates section. (Image courtesy of NASA/Kepler mission/Wendy Stenzel.)

an Earth-trailing heliocentric orbit until May 2013 (Figure 5.8). With its larger telescope and field of view, Kepler is potentially able to detect more exoplanets than CoRoT. Indeed, after three years in operation (by December 2012), Kepler has detected about 100 confirmed exoplanets and over 2000 candidate exoplanets, which has very strong implications in terms of exoplanet statistics.

Class III/IV habitats may be found by measuring the mean densities of discovered super-Earth-type exoplanets, and this possibility has been studied for present-day transit missions in space such as CoRoT (CNES) and Kepler (NASA) in combination with ground-based Doppler velocimetry measurements from HARPS (ESO) and possible future instruments. Particular attention has been paid to the sources of uncertainty in the planetary density which are related to uncertainties in the mass determination by radial velocity measurements, stellar radius determination, and photometric measurement during the transits. With the instruments currently available, the accuracy of radial velocity measurements is the main uncertainty and limiting factor for expected detections by CoRoT and Kepler.

As a consequence, the determination of the nature of such planets only seems possible if their host star is bright enough. It has been shown that if each star in the CoRoT field had a 6–10 Earth-mass planet at about 0.10 AU, the number of detections with CoRoT would be several hundred. On the other hand the absence of any detection would indicate that fewer than 1 per cent of stars have such planets.

However, a definite answer to these open questions depends on the photometric precision of CoRoT and Kepler. New generations of radial velocity instruments are necessary, which can make accurate measurements on faint stars. In that case, the identification of possible ocean planets could be done on a significantly larger stellar sample.

### 5.1.5    *Gravitational microlensing*

In parallel with the velocimetry and transit techniques, an original method has been developed for detecting exoplanets: gravitational microlensing. The principle uses an application of the theory of General Relativity developed by Albert Einstein (1879–1955), according to which the light rays are curved in the vicinity of a very massive object. The first evidence for this phenomenon was given in 1917, as the light coming from distant stars, close to the position of the Sun in the sky, was observed at the time of a solar eclipse. This method was later used to search for 'dark matter' in an attempt to solve the mystery of the missing matter in the Universe. If a black massive object transits in front of a distant source, the light of the object is deflected around the black object and converges on a single point to show a transient amplification of its signal (a 'flash') of a certain well-defined pattern at the exact time of the alignment. Ground-based networks were set up to search for gravitational microlensing phenomena, which would have shown evidence for massive objects. In particular, the MACHO (Massive Compact Halo Objects) project made systematic observations of background objects in a nearby galaxy to try to detect massive compact halo objects. Some events were observed, but the total number of such dark massive objects appeared to be small compared with the missing matter of the Universe.

The search for exoplanets using microlensing is derived from this technique. When a star passes in front of a bright distant object, the light of the distant object is amplified owing to the mass of the transiting object: this is the microlensing effect. If a planet is present around the star, a narrow peak appears in the wing of the stellar curve, corresponding to the time of the alignment of the planet and the distant source. A few planets, including a small one of less than 10 terrestrial masses, have been detected by this technique. Its limitation is that the phenomenon, which requires a specific geometry, is purely random and thus cannot be predicted or repeated. A few ground-based networks, in particular PLANET (Probing Lensing Anomalies NETwork), MOA (Microlensing Observations for Astrophysics) and OGLE (Optical Gravitational Lensing Experiment) have been designed for detecting lensing or microlensing events. They have indeed detected exoplanets, but in many cases, those were transiting exoplanets.

## 5.1.6  Indirect methods: what do they tell us?

Over the past decades, several indirect methods have been used to detect exoplanets. Two of them, astrometry and velocimetry, are designed for measuring the motion of the host star with respect to the centre of gravity of the star–planet system. Astrometry is sensitive to motions in the celestial sphere, i.e. perpendicular to the star–observer axis. In contrast, the velocimetry technique measures variations of the star along the star–observer axis. The two methods are thus complementary. Astrometry can in principle detect all planetary systems: the motion of the star on the sky is an ellipse, ranging from a full circle (if the planetary system is seen from above) to a single line (if the observer is in the orbital plane of the planet). In contrast, the velocimetry method is not sensitive to planetary systems as seen from above, because the radial velocity of the planet is zero in this case. In addition to the period of the planet, the physical quantity retrieved from these observations is, for astrometry, the mass $M$ of the planet, and, for velocimetry, the quantity $M \sin i$, with $i$ being the inclination of the system ($i = 0$ corresponding to a system as seen from its pole).

As mentioned above, astrometry was the first method to be attempted for detecting exoplanets, but its sensitivity was not sufficient at the time of these early observations. New attempts will be made in the coming decade using higher-performing instruments on large telescopes, at the VLT (the Very Large Telescope in Chile) in particular. But the most promising perspective will come from the space mission GAIA, to be launched by ESA in 2013, which will measure the positions and motions of billion of stars. Both astrometry and velocimetry are especially well adapted for massive exoplanets; the detection is made easier for planets close to their star as the time needed for a revolution is shorter. Fortunately, as we have seen, such exoplanets are quite common around nearby stars.

The transit method leads to the determination of the planet's diameter and its orbit's inclination (provided it is close to 90°) and thus, with the information on the mass coming from the velocimetry, allows us to determine the mean density of the body. The method is very well adapted to large objects close to their stars (again the majority of the observed exoplanets). The limitation comes from the peculiar geometry needed for the observation, as the planetary system must be seen edge-on. In the case of giant exoplanets very close to their star (with a 4-day period), the probability of transit is 10 per cent. Another great advantage of the transit method, as will be seen below, is that, in the case of bright stars, it can allow characterization of the exoplanet's atmosphere, either during primary transit (when the planet passes in front of the star) or during secondary transits (when the planet passes behind the star).

The microlensing technique provides information about the mass and the projected distance of the exoplanet. It is very well adapted to small objects distant from their star. In contrast to previous techniques, it allows for the detection of exoplanets around distant stars, such as those located in the halo of our Galaxy, and thus can provide us in the future with statistical information on the exoplanet population. The limitation is that these unpredictable and unrepeatable observations cannot provide information on the physical nature of the exoplanets.

For completion, we should add the pulsar timing method, which was the first one to lead to the detection of an exoplanet in 1992. This method is extremely sensitive to low-mass planets (two of the detected planets are similar to Mercury and Venus in size) but is limited to very few stars, the pulsars. Only a few planets have been detected by this method since the first discovery 20 years ago.

## 5.1.7 *Imaging exoplanets at last*

As pointed out earlier, the great difficulty in imaging exoplanets lies in the high flux contrast between the star and the planet in the visible range. As seen from 10 parsec (about 30 light-years) a Jupiter-type exoplanet would have a visible light flux $10^7$ times lower than its star (assuming this star is a solar-type star), and it would be located at 0.5 arcsec from its star; this distance corresponds typically to the best image quality which can be expected for a ground-based observatory. However, the situation gets better if we move toward infrared wavelengths, as the flux contrast depends upon the intrinsic temperature of the planet (see Box 5.3). In the case of a giant exoplanet close to its star (a so-called 'hot Jupiter'), at 10 μm, the flux contrast between the star and the planet would become $10^5$, i.e. 100 times better than in the visible; but the angular distance between the star and the planet would be only 5 milliarcsec! For an Earth-like planet at 1 AU from its star, the flux contrast would be about $10^6$, and the angular separation (the angular distance between two objects as perceived by an observer) would be 0.1 arcsec.

Observing exoplanets directly, then, means finding the best compromise between the flux contrast (which depends upon the effective temperature of the planet and its star) and the angular distance between the two objects. Two conditions should be privileged: (1) choose a low-mass star with a low effective temperature; (2) look for distant exoplanets to improve the angular separation.

BOX 5.3    **The radiation of stars and planets**

Spectroscopy of a stellar or planetary body is a powerful tool for measuring its temperature and determining the constituents of its atmosphere and/or surface. In the case of stars, the general shape of their spectrum allows us to determine their equilibrium temperature, i.e. the temperature of the black body which best fits this spectrum. Solar-type stars (of G type) have a maximum which peaks at 0.5 μm. More massive stars (of O, B and A types) peak at shorter wavelengths whereas smaller ones (K and M types) have their maximum emission shifted toward the infrared.

   A planet orbiting around a star receives the stellar radiation which may be either reflected or scattered back at the same wavelength, or absorbed and converted into thermal emission, which peaks at longer wavelengths. The equilibrium temperature of the planet is defined as the temperature of the black body that would emit the same flux. The equilibrium temperature $T_e$ of a planet can be calculated as follows:

$$[F^*/D^2](1 - a) = 4 \, \sigma T_e^4$$

where $F^*$ is the flux of the star, $D$ is the distance of the planet to the star, $a$ is the albedo (the fraction of stellar flux reflected by the planet) and $\sigma$ is a factor known as the Stefan constant. This formula applies if the planet (like the Earth and the giant planets of the Solar System) is in rapid rotation: in this case the flux received on the dayside hemisphere is re-radiated in all directions. In the case of a slow-rotating planet (like Venus, Mercury or all exoplanets close to their stars which are tidally locked to them), the formula becomes:

$$[F^*/D^2](1 - a) = 2 \, \sigma T_e^4$$

   In the case of Solar System objects, the albedo is typically 0.3 for planets; it generally ranges from 0.1 to 0.6 in the case of asteroids and outer satellites (with Enceladus being exceptionally bright with an albedo of 0.9). The albedo of TNOs is typically 0.10 while comets are usually darker (a few per cent). In the case of exoplanets, low albedos (typically 0.03) have been measured in the case of hot gaseous Jupiters, probably as an effect of scattering. Equilibrium temperatures of hot Jupiters typically range between 1200 and 2500 K, depending on their distance to the star and the spectral type of this star.

BOX 5.3  **(cont.)**

It should be mentioned that the equilibrium temperature of an exoplanet may differ from its true effective temperature (the one that corresponds to its actual thermal emission): this is the case for giant planets with an internal energy source, or terrestrial planets with a greenhouse effect. But the definition of the equilibrium temperature allows us to determine, to first order, the type of spectrum that can be expected for the exoplanet.

Like any Solar System object, the spectrum of an exoplanet is a combination of the reflected/scattered stellar component and the thermal emission. In the case of Solar System planets, the crossover between the two components is at about 3 μm for the terrestrial planets and beyond 10 μm for Uranus and Neptune. In the case of hot Jupiters, the thermal emission usually dominates beyond 1 μm.

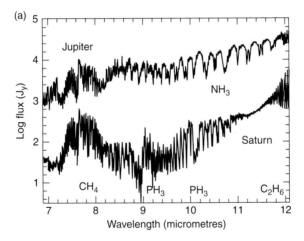

FIGURE BOX 5.3(a)  The spectrum of Jupiter and Saturn, measured by the short-wavelength spectrometer of the Infrared Satellite Observatory (ISO) between 7 and 12 μm. This part of the spectrum corresponds to thermal emission. It can be seen that molecular signatures appear either in emission ($CH_4$ at 7.7 μm, $C_2H_6$ at 12 μm) or in absorption ($NH_3$ on Jupiter at 10.5 μm, $PH_3$ on Saturn at 9 μm and 10.2 μm), depending where the bands are formed in the atmosphere. (Adapted from Encrenaz *et al.*, 1999.)

BOX 5.3    **(cont.)**

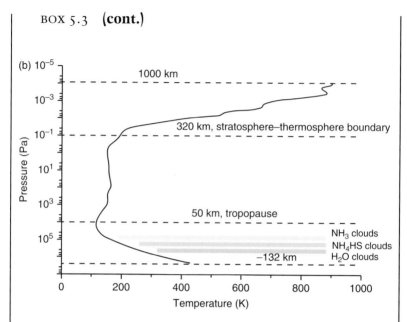

FIGURE BOX 5.3(b)  The atmospheres of both Jupiter (shown in this figure) and Saturn exhibit a temperature inversion above the tropopause, at a pressure level of $10^4$ Pa (100 mbar), owing to the presence of a stratosphere induced by methane photochemistry. As a result, molecular bands formed in the stratosphere are shown in emission, while the tropospheric signatures appear in absorption, as shown in panel (a). (Redrawn from image courtesy of Wikipedia Commons.)

The distinction between reflected and thermal emission is important in understanding the spectrum of an exoplanet, because the mechanism of spectral line formation is not the same in the two cases. In the reflected starlight component, the spectral signatures of the exoplanet's atmospheric constituents are observed in absorption superimposed on the stellar continuum; they allow the identification of the species and the measurement of its column density, integrated over the line of sight. In contrast, in the thermal regime, the radiation probes the atmospheric level where opacity is equal to unity. In the case of a weak molecular band, deep tropospheric levels are probed, whereas in a strong molecular band ($CH_4$ at 3.3 and 7.7 μm, for instance), the radiation probes the stratosphere.

An important parameter is the thermal profile which determines the shape of the molecular signatures. In the case of Mars and Venus, which

BOX 5.3   **(cont.)**

have no stratosphere, the molecular lines always appear in absorption. In contrast, in the case of the giant planets and Titan, a temperature inversion occurs in the stratosphere. As a result, tropospheric signatures appear in absorption whereas stratospheric signatures are seen in emission. In the case of an exoplanet, as no a priori information is known about the thermal profile, the interpretation of thermal spectra may be ambiguous. Ideally, one needs to determine simultaneously the thermal profile and the vertical distributions of the atmospheric compounds. This may be done if both the reflected and thermal components are measured, and/or if different bands of a given species, with different intensities, are simultaneously measured, allowing atmospheric 'sounding' at different altitudes. In both cases, a large spectral interval has to be covered.

The near-infrared spectral interval is of special interest for imaging exoplanets because, in addition to the contrast gain with respect to the optical range, high-precision photometry can be achieved, using coronagraphy and adaptive optics (see Subsection 2.4.3). This is how the first exoplanet 2M1217 b was imaged in 2005 at the VLT, by Gael Chauvin and his team. Its host, 2M1217, is a low-mass star identified by the 2-MASS infrared star survey, with an effective temperature of about 3500 K. The planet, with a mass of 5 Jovian masses, is located at 55 AU from its host star (Figure 5.9). Since 2005, about 30 exoplanets have been imaged directly, including two multiple planetary systems. Recently, a three-planet system was imaged around HR8799, a massive main-sequence star five times as bright as the Sun, located at 39 parsec from the Sun.

## 5.1.8   *Detecting exoplanets through their radio emission*

Another direct detection method has to be considered, at the other end of the electromagnetic spectrum. There is one spectral region where the flux contrast between the exoplanet and its star may be very

FIGURE 5.9  First detection of an exoplanet by direct imaging. The figure is a composite image of the brown dwarf 2M1207 and its companion in H (blue), K (green) and L' (red) filters. (From Chauvin *et al.*, 2004.)

favourable: this is the radio range, in particular at decametric wavelengths (wavelengths between 10 and 100 metres). An advantage is that in this spectral range, the Solar System planets with strong magnetic fields (i.e. Earth and the giant planets) produce auroral phenomena as a result of the interaction between their magnetic fields and the charged particles coming from the solar wind. In the case of magnetized giant exoplanets, the intensity of the decametric emission associated with these aurorae might be comparable to that of the star itself.

The detection of such events, however, is expected to be limited by two sources of noise: the fluctuations of the sky background coming from the Galaxy, and the parasitic signals coming from the Earth itself, in particular those of human origin. Nevertheless, observing campaigns have been conducted using single dish antennae and more recently LOFAR (the Low Frequency Array).

## 5.2 THE EXOPLANETARY ZOO

In mid 2013, about 900 exoplanets have been identified in 700 planetary systems. A first question that must be asked, especially for astrobiological purposes, is: how common are planetary systems? More than 15 years after the discovery of 51 Peg b, radial velocity programmes have given us some statistics. From a sample of 1200 FGKM stars, it appears that about 7 per cent host at least one giant planet of at least half a Jovian mass, located within 5 AU of their host star. Among these, about 15 per cent are 'hot Jupiters', i.e. giant exoplanets orbiting around their primaries in less than 10 days. The frequency of giant exoplanets seems to increase with the metallicity of the star (Box 5.4), as well as with the stellar mass and

---

BOX 5.4 **The metallicity of the stars**

Metallicity, written as [Fe/H], is a ratio which allows astronomers to categorize stars as a function of the amount of iron versus hydrogen they contain. The term 'metal' here refers to anything heavier than the gases hydrogen and helium. The content in iron and hydrogen of a star is defined by its absorption signatures present in its spectrum, normalized to the equivalent values in our Sun. Because the metallicity is expressed in logarithmic values, [Fe/H] = $-1$ means that this particular star has a tenth as much 'metal' as the Sun does, and [Fe/H] = $+1$ indicates 10 times the solar content.

Nucleosynthesis in the Big Bang theory predicts the formation of essentially just hydrogen and helium in the beginning (with small amounts of lithium, deuterium and beryllium) so that all the heavier elements we see today in the Universe (the 'metals') must have been produced in the interior of the stars themselves. When the star reaches the final phase of its life it re-injects some of these 'metals' into the interstellar medium, thus enriching the environment in which the next generation of stars are born so that the new stars in turn become enriched with their ancestors' material. By measuring this enrichment we find implications for the age of the stars, because those with a lower metal content (smaller [Fe/H]) are older than those with a higher metal inventory (larger [Fe/H]).

BOX 5.4   **(cont.)**

The radial velocity planet surveys have offered us information on the planet–star metallicity relation, suggested as early as 1997 by Guillermo Gonzalez who remarked that the stars hosting planets had a higher metallicity than others without planetary systems (Gonzalez 1997). Since then, observations and models have reported a consensus on the fact that independently of their mass, all stars that have giant planets orbiting around them tend to be metal-rich. But in addition it would appear that giant planets tend to favour more massive stars as their hosts. These findings, if confirmed now that we know of more and more exoplanetary systems, would tend to prove that giant planet formation is the result of the core-accretion mechanism. In that theory, a giant gas planet like our Jupiter forms by first assembling a core of icy/rocky/metallic material. The core accretes more efficiently and quickly in the presence of enough nebular gas, which is found mainly in massive protoplanetary disks. This effect explains the connection between giant exoplanets and stellar mass. It has also been shown that the presence of larger surface densities of solids in the disk is a positive factor in the formation of giant planets, thus pointing to the planet–stellar metallicity relation. A metallicity correlation also exists for low-mass planets orbiting lower-mass stars.

the mass of the stellar disk. Following the statistics of the Kepler mission, it now appears that the probability for a star to host a planet would be over 50%.

### 5.2.1   *The brown dwarf desert*

What is the mass distribution of the exoplanets? An early discovery was the evidence for a bimodal or twin-peaked distribution, with a population of substellar objects with masses above 80 Jupiter masses (i.e. 0.08 solar mass), and a 'planetary population' with masses of a Jovian mass or less (i.e. less than 0.01 solar mass; Figure 5.10). The region between these two populations is called the 'brown dwarf desert'.

This dichotomy illustrates the different formation scenarios of the two classes of objects. The limit of 0.08 solar mass corresponds to the threshold beyond which nuclear reactions can take place in the core of the object. Above this limit, the object is a star; below this limit,

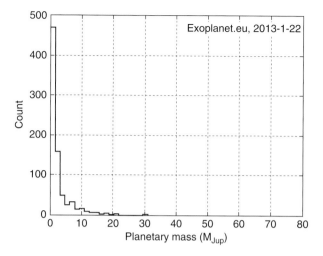

FIGURE 5.10 Histogram of exoplanets as a function of their mass. The 'brown dwarf desert' extends beyond a few tens of Jovian masses. The population of heavier substellar objects is not shown here. (Redrawn from *The Extrasolar Planets Encyclopedia*; J. Schneider, Observatoire de Paris; http://exoplanet.eu.)

down to 0.01 solar mass, the object may still harbour nuclear reactions due to deuterium fusion. The object is then called a brown dwarf; its radiation is only $10^{-4}$ times the solar one, but it lasts longer than the Sun. Below 0.01 solar mass, or 13 Jovian masses, one enters the realm of the planets. As discussed above, their formation mechanism, based upon accretion within a protoplanetary disk, is completely different from the stellar formation, which explains the lack of objects between the two populations.

The first exoplanets discovered were very massive. This was partly an observational bias, because they were the easiest ones to detect, as they had a strong effect on the velocity of their host star. Over 15 years after the first discoveries, it now appears that low-mass exoplanets are becoming more and more numerous. They are, in particular, found within multiple systems, and suggest that a large population of low-mass objects is to be discovered in the coming years. Using both transit and velocimetry techniques, we now know about 20

Earth-type or 'super-Earth' exoplanets – presumably rocky exoplanets – of masses less than 10 terrestrial masses. Many of them are close to their star and/or belong to multiple systems, like those around GJ 581, Kepler 11, Kepler 20 or Kepler 33 (see below).

### 5.2.2    Exoplanets close to their stars

The first exoplanets discovered were also very close to their host stars. At that time, this was considered to be another observational bias: short-period objects were easier to detect. This trend has been con-firmed, however, over the past years: there is indeed a peak at 3–10 days in the period of exoplanets, corresponding typically to a distance of 0.05–0.1 AU. As mentioned above, these objects are believed to have formed at distances further from their stars and to have migrated inward down to a distance of 0.05 AU; the mechanism responsible for stopping this migration is still a matter of debate. Another peak in the distance distribution might show up between 0.3 and 3 AU; at larger distances, more long-term observations are required.

A correlation also seems to exist between the period of a planet and its mass (Figure 5.11): close-in, short-period planets appear to be less massive than the distant, longer-period ones (although hot Jupiters were among the earliest discoveries, as a result of the observational bias as described before).

### 5.2.3    Exoplanets on eccentric orbits

An unexpected result of the exoplanet exploration is the values of the eccentricities of many objects. While close-in planets preferentially rotate on circular orbits, exoplanets with periods above a few days exhibit a wide range of eccentricities, with a median value of about 0.3. Tidal circular-ization is expected to take place for close-in planets, but the high eccen-tricity of the more distant ones could be the result of gravitational interactions in multiple systems, or interactions between the planet and the disk.

One might thus wonder: are there Solar System analogues among exoplanets? There does indeed appear to be a small sample of

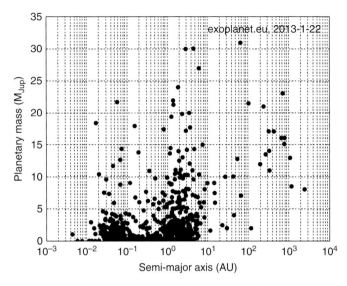

FIGURE 5.11 Planetary masses as a function of the distance to their host star. The diagram shows that many giant exoplanets orbit in the vicinity of their host stars. (Redrawn from *The Extrasolar Planets Encyclopedia*.)

such planetary analogues. A Jupiter analogue of 3.8 Jovian masses, with an eccentricity of 0.06, can be found in the 55 CnC planetary system, at 5.8 AU from the 0.94-solar-mass 55 CnC star. Another example can be found around the 0.88-solar-mass star HD 154345: the object, with a minimum mass of 0.94 Jovian mass, lies at a distance of 4.2 AU with an eccentricity of 0.16. As long-term observing campaigns have now been accumulating data for more than a decade, there is little doubt that more analogues for the Solar System's giant planets will be found in the future.

## 5.2.4 Many exotic objects

The detection of exoplanets very close to their host stars has naturally led to the discovery of new classes of hot objects, very different from the Solar System planets. They are of two kinds: either they are giant planets, called 'hot Jupiters', or they are small and dense objects.

Dozens of hot Jupiters have been discovered. The most famous is HD 209458b (sometimes also named Osiris), already mentioned, which was the first exoplanet detected by transit. The system is located at 47 parsec from the Sun. The planet, 0.045 AU from its star, has a revolution period of 3.5 days. It has a mass of 0.69 Jovian masses, but its radius is as much as 1.3 times that of Jupiter. As a result, its density is only 0.4 g cm$^{-3}$, i.e. one-third of Jupiter's. Spectroscopic measurements in the visible and ultraviolet range have shown the presence of sodium and hydrogen atoms, respectively, in an area much wider than the radius itself, which implies that the atmosphere is evaporating, owing to the strong radiation pressure of the nearby star. The evaporation rate has been estimated to be 10 000 tonnes per second. At such a rate, the planet would probably lose about 10 per cent of its mass within its lifetime.

We now know many hot Jupiters; some are closer to their stars, others are less massive than HD 209458b. In some cases, one could imagine that the whole atmosphere of the planet might be swept away within the lifetime of the star. In this case, the residual would be a small, rocky and very dense object. This leads us to the second class of exotic objects found in the immediate vicinity of the stars: the small, dense planets, possibly made of a metallic and melted silicate core. In this category we find CoRoT-7 b, Kepler-10 b, Kepler-11 b and 55 CnCe. Like the hot Jupiters, these objects, expected to be phase-locked, probably exhibit very different hemispheres on the dayside and on the nightside. Located at 150 parsec, CoRoT-7 b (Figure 5.12) is a planet of about 5 terrestrial masses, with a radius of less than 2 terrestrial radii, orbiting at 0.01 AU from its star in less than a day. Kepler-10 b and 55 CnCe have comparable characteristics. The possible structure of such objects is much discussed in the literature. However, these exotic objects probably hold little interest in terms of exobiology or habitability, as the closeness to the primary star means that the temperatures are formidable.

## 5.2.5    *A large number of multiple systems*

Multiple planetary systems appear to be common around nearby stars (Figure 5.13). About 100 multiple systems have been discovered by

FIGURE 5.12 The exoplanet CoRoT-7 b, an extreme example of a small exoplanet located in the immediate vicinity of its host star; artist's impression. For colour version, see plates section. (Image © ESO/ L. Calçada.)

FIGURE 5.13 The Kepler 11 planetary system with six planets around their star, discovered by the satellite in 2011; artist's impression. For colour version, see plates section. (Image courtesy of NASA/T. Pyle.)

velocimetry and/or transit. From a theoretical point of view, their existence is not a surprise. The nucleation model of planet formation supports, as in the case of the Solar System, the accretion of multiple embryos. The migration mechanism, proposed to explain the presence of giant planets close to their stars, can probably also explain the frequent orbital distribution of planets in or near mean motion  resonance (i.e. when the planets' revolution periods are in simple ratios), and may also be responsible for the large range of observed eccentricities. Resonant systems can be stable and the planets remain in resonance indefinitely; in other cases, the bodies exchange momentum and the planetary orbits can move inward or outward, with a planet being expelled or absorbed within the star. Such calculations have an important implication for exobiology: indeed, dynamical simulations testing the stability of resonant systems can allow us to determine whether the presence of Earth-like planet is possible in these environments.

### 5.2.6  Planets around multiple stars

Even more exotic are the exoplanets orbiting a binary star (Figure 5.14). They can be either circumstellar objects (orbiting around one of the stars) or circumbinary (surrounding both stars). Several tens of such objects have been discovered, mostly giant planets orbiting the primary star. Few objects, especially few giants, have been found for binary separations smaller than 100 AU; this observation is consistent with models predicting that the giant planet formation is inhibited by the presence of a nearby star. An exciting discovery was the detection of PH1 (Planet Hunter-1), a system of two planets (KIC 4862625 A ab) orbiting a star belonging to a binary system.

### 5.2.7  Which candidates are most likely to be habitable?

If we are looking for life on an exoplanet, what should we be searching for? By simple analogy with what we know about possible favourable conditions for life, we can set up a list of criteria for the properties of 'habitable' exoplanets. We have seen that four factors are believed to be

FIGURE 5.14 Artist's impression of an exoplanet orbiting both stars in a binary star system. For colour version, see plates section. (Image credit: original art by T. Encrenaz.)

necessary for life: (1) the presence of liquid water; (2) the presence of carbon; (3) an energy source; and (4) the stability of the system.

First, the planet should have a surface to sustain a possible water ocean. This favours the choice of Earth-like planets or super-Earths (Figure 5.15), but does not exclude solid satellites of giant exoplanets, which would be the analogues of our outer satellites; such objects remain to be discovered in the future.

The presence of liquid water sets a constraint on the temperature of the exoplanet. The temperature should be between 0 °C and a few hundred °C, depending on the pressure. The temperature of the planet results from a balance between the radiation flux absorbed from the star and the flux emitted by the planet itself. The flux depends upon the distance between star and planet, but also on the spectral type of the star and its mass. As we have seen in Chapter 2, astronomers have defined the principle of the 'habitability zone' or HZ, which describes the distance range where water may be in liquid form for some time. For a solar-type star, this zone (which of course includes the Earth) ranges from about 0.7 to 1.5 AU. If the star is brighter, the HZ is shifted

FIGURE 5.15 Artist's impression of an Earth-like world in a different planetary system. For colour version, see plates section. (Image credit: original art by R. Poux.)

outwards; if it is fainter, it is reduced to distances closer to the star. An interesting case is the M-dwarf population which represents about 90 per cent of the whole stellar population. In that case, the distance of the HZ is only 0.1 AU. This is very favourable for observations, because the short revolution period makes the transit easier to detect and to observe. As will be seen below, transit observations are important not only for detecting planets, but also for characterizing their atmospheres. We thus have a key to probe the atmospheres of the transiting exoplanets: this is the second revolution in the exploration of exoplanets.

In particular, the system around the local red dwarf Gliese 581 (located at 20.3 light years from the Sun, spectral type M3V) has attracted a lot of interest over the past decade owing to the six low-mass exoplanets discovered around it by Swiss astronomers Stéphane Udry, Michel Mayor and colleagues (Udry *et al.*, 2007). One of these, GJ581g, was proposed to be located in the habitable zone, but the very existence of this planet remains controversial. On the other hand, in the same system, exoplanet GJ581d, discovered in 2007, is a low-mass planet (2–10 Earth masses) that has been confirmed as one of the few super-Earths known today. In spite of its large distance to its primary star which would allow it to receive only a third of the insolation that arrives on Mars, and thanks to the possible presence of an atmosphere including large amounts of $CO_2$, a greenhouse effect might occur there which would cause the temperatures to rise and allow for liquid water to exist on the surface. However, it is also quite possible that this planet is locked in tidal resonance with its star; and this, as we have seen before, is clearly a showstopper for habitability in many cases owing to the permanent night and cold conditions on one side of the object. Modelling of the planet's climate in the presence of an atmosphere (Wordsworth *et al.*, 2011) has shown that in many different cases the habitable conditions would be preserved. In order to better constrain such models, observational tests are needed in the future.

For terrestrial planets found in the insolation habitable zone (IHZ) of low-mass stars (LMS), tidal effects caused by the planet's spin orientation and rate have severe effects on their habitability. First, tidal erosion of the obliquities ('tilt erosion') of such planets occurs much faster than life emerged on Earth. Second, tidal heating is significant for Earth-like planets in the IHZ of stars with less than 0.25 solar masses. Studies of tidal processes on the super-Earth Gl581d and the super-Earth candidate Gl581g show that Gl581d's obliquity is eroded to nil and its rotation period reaches its equilibrium state of half its orbital period in less than 0.1 Gyr. As we have seen, obliquity tides affect the habitable potential of a planet and need to be taken into account for studies of surface processes. Tidal surface heating on Gl581g, for example, would remain at levels lower than $150 \text{ mWm}^{-2}$ as long as its eccentricity is smaller than 0.3.

In addition, it has been shown (Wordsworth, 2012) that for some exoplanets there can be periods of time in their histories when conditions for life may have been available. Although major uncertainties remain in planetary formation theory, it is generally believed that planets of between 1 and 20 Earth masses can form surrounded by primordial hydrogen envelopes due either to capture of the gas from the solar nebula or to outgassing during accretion. Studies of the impact of hydrogen greenhouse warming for such exoplanets that are closer to their star, with lower equilibrium temperatures than Earth, and which can therefore more easily lose their hydrogen atmospheres entirely, have shown that they could go through transient periods where the hydrogen in the atmosphere would provide sufficient greenhouse warming (through $H_2$–$H_2$ collision-induced absorption), and therefore they might witness the presence of oceans on their surfaces, at least for some time. Even if the hydrogen layers were later lost to space, photochemical byproducts such as methane or carbon dioxide might have been formed. These could then take over to sustain the greenhouse effect, so that transient periods with habitable conditions would always be an option, although for how long largely depends on the star's type (more favourable for M-class ones), the size of the planet and its $H_2$ inventory. If the conditions in such a reducing atmosphere allow it, then perhaps some basic life may have had the chance to emerge. For that to happen, it would have to fight against the unfavourable conditions of an environment dominated by hydrogen-bearing species, where the energy to allow redox reactions to happen would come from gaseous volcanic emission of photochemical processes, for instance. We know that prebiotic molecules readily form in the reducing chemistry of hydrogen-rich atmospheres, so it is not ruled out that life might eventually find a way to appear on these planets during these transient periods, but whether it might survive depends on its ability to adapt to (or to act upon and change) its irrevocably cooling environment. If this is not the case, then the existence of life on such planets would be such a short-lived phenomenon that it would not make sense to search for it.

So, what are the next steps in future exploration of extrasolar habitats? One of the major goals in the short term is the search for exomoons, for instance by transit photometry, by the so-called Rossiter–McLaughlin effect (a spectroscopic signature of the transit curve associated with the Doppler shift, which may be used to infer the retrograde motion of some exoplanets), or by planet–moon transits, for reasons we have explained above. Recently, the Kepler mission has dedicated time to search for exomoons within the 'Hunt for Exomoons with Kepler' (HEK) project, and it seems that success is just around the corner. Scientists such as René Heller and Rory Barnes have worked together on defining what would make an exomoon habitable and have come up with the idea of the 'habitable edge'. This is the orbit around the planet beyond which the satellite would not be submitted to the runaway greenhouse effect, which we suppose is what left Venus without any water after it evaporated because of the increasing solar heat. If exomoons exist, around the large gas giants ('hot Jupiters') we find around some distant stars, then they might be able to have an atmosphere and develop a magnetic field to protect it, as well as find some energy sources to create habitable conditions.

Our major efforts today focus on transiting planets because some of their characteristics such as obliquity can be inferred from observations, in particular the Rossiter–McLaughlin effect just mentioned. Knowing the obliquity of the planets is essential to determine tidal heating and atmospheric conditions on a planet, which leads to an evaluation of the habitable planets in the IHZ around their host stars. Thus, transiting planets are the most promising targets for near-term searches for habitats.

## 5.3 FROM DETECTION TO CHARACTERIZATION

So far we have concentrated on the detection of exoplanets. The various methods used have given us information about a few basic parameters: the distance to their host star (which leads to an estimate of their equilibrium temperature using the spectral type and the mass of the host star), their eccentricity, their mass (or a lower limit of it in the case of objects detected by velocimetry), and in addition, in the case of transiting planets, their diameter (which leads to a determination of their densities). Using these basic orbital and physical properties, we have been able to infer statistical properties of the exoplanets' populations.

As described above, we have also identified extremely exotic objects, either very dense or very light, close to their stars or far away, or on highly elliptical orbits. We have also identified a surprisingly high number of multiple planetary systems, which are excellent targets for testing dynamical models of planetary systems. But all this information does not tell us the nature of their atmospheres: are they like the Earth or more like the giant planets that we know? This question opens a new field of research, and its answer requires the development of new tools. Over the past few years, these tools have started to become available: they consist in observing the atmospheres of transiting exoplanets.

### 5.3.1   Primary transits

We have described primary transits, when the planet passes in front of the star, as a powerful method for detecting exoplanets, but actually they offer more than that. If, instead of photometry, spectroscopy is achieved during the primary transit (Figure 5.16), information can be retrieved about the composition of the exoplanet's atmosphere. Indeed, the radius of the exoplanet now becomes a function of wavelength. Let us suppose that a molecule, for example methane, shows absorption signatures in the spectral range of the observation (typically the near-infrared range). If methane is present in large quantities in this atmosphere, the radius measured at the wavelength of methane absorption will be larger than at other wavelengths. Using some assumption about the temperature and pressure of the planet's atmosphere, it is possible to infer information about the total content of methane along the line of sight. If different molecules are observed, relative abundances can be inferred. Calculations show that the amount of light crossing the planet's atmosphere depends upon its scale height, by which we mean the distance in which the pressure is divided by a factor $e$ (the base of the natural logarithm). It can be shown that the scale height is proportional to the temperature, but inversely proportional to the mean molecular weight of the atmosphere and the gravity of the planet. As a consequence, the

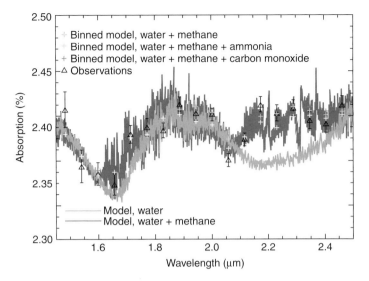

FIGURE 5.16 An example of primary transit observations of HD 189733b. As seen in this figure, models with water vapour only cannot fit the data around 2.2–2.4 μm, while a model including both water vapour and methane provides good agreement with the observations. (Redrawn from Swain *et al.*, 2008.)

hotter and less dense the planet's atmosphere is, the easier the detection. Hot Jupiters, having hydrogen-rich hot atmospheres and transiting close to their stars, are especially favourable targets. Their scale height can reach several hundred kilometres, while the terrestrial one is only 8 km.

First detections of atmospheric species using the primary transit method were obtained in the early 2000s from ultraviolet and visible observations, mostly using the Hubble Space Telescope (HST). These spectral ranges typically probe upper atmospheres where atmospheric species react with the stellar flux, and long column densities (the total number of atoms along the line of sight) can be measured. The first species detected by this method were atoms such as O, H, C and Na; targets were limited to a few of the brightest hot Jupiters, in particular HR 209458b (the first detected transiting planet). Later, observations were performed, with the HST and also from the ground, in the near-infrared where molecular signatures, coming from lower atmospheric levels, can be probed. Detections of water vapour and methane were reported by this method. Because observers are working at the limit of feasibility, contradictory results have sometimes been reported, leading to some controversy in the scientific community. Ten years after the first detections, it now appears that primary transit spectroscopy is a powerful method for characterizing the upper atmospheres of exoplanets. Mostly limited to hot Jupiters at present, it should be easily applicable to other classes of exoplanets in the future.

It is interesting to note that primary transit spectroscopy probes the exoplanet's atmosphere at the terminator, the transition between dayside and nightside. In the case of hot Jupiters, the proximity to their host star implies that they are phase-locked, i.e. they always turn the same face to their star, just like the Moon does with the Earth. The dayside and the nightside are expected to be very different. As a result, the terminator may be a region of instability, also different from the day and night sides.

## 5.3.2   *Secondary transits*

When an exoplanet transits across its star, the light curve of the star during a full revolution period does not contain only the flux depletion corresponding to the primary transit. When the exoplanet passes behind the star, a slight decrease of the stellar flux is observed, smaller than the dip corresponding to the primary transit. This time, it corresponds to the emission of the planet itself. This is actually a truly direct detection method: the flux of the exoplanet is obtained by subtraction between the stellar flux just before or after the secondary transit, and the stellar flux alone during transit, when the planet is behind it. The flux difference observed during a secondary transit is usually weaker than the one corresponding to the primary. This difference directly depends upon the flux ratios between the star and the exoplanets; it is thus advantageous to use infrared wavelengths, as much as possible in the far-infrared.

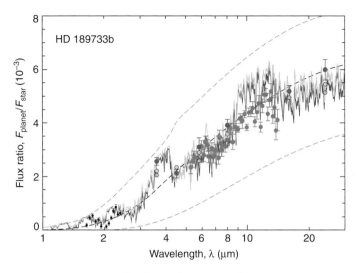

FIGURE 5.17  An example of secondary transit observations. Data are from the HST (NICMOS) at short wavelengths, and from Spitzer (IRS and MPIS) in the thermal infrared. Comparison is made with photochemical models. (Redrawn from Moses *et al.*, 2013.)

Secondary transit spectroscopy (Figure 5.17) has enormously benefited from the use of a space observatory – Spitzer – which turned out to be operating at the right time in the right place. Spitzer is an infrared telescope launched in space by NASA in 2003, located on a heliocentric orbit, which operated over the whole infrared range until May 1999, when it lost its liquid helium; it is still operating now in the near-infrared range, for which detectors do not require active cooling. Spitzer was mostly designed for cosmology and extragalactic studies, but turned out to be perfectly adapted to the study of exoplanet transits. By choosing the few brightest hot Jupiters and following the light curve of their host stars along the revolution period of the planet in the near-infrared and far-infrared range (up to about 30 μm), Spitzer was able to identify atmospheric species: water vapour and methane again, but also carbon monoxide, carbon dioxide and ammonia. Again, some of the results are still controversial. However, they illustrate the great potential of this technique, which will no doubt flourish in the coming decade.

### 5.3.3 How to search for life in exoplanetary atmospheres

Spectroscopy of exoplanets, in the visible and infrared range, offers a promising way of probing their atmospheres for investigating the presence of possible traces of life. As mentioned above, the most favourable targets will be searched for in the 'habitable zone', where the temperature is expected to allow water to be in liquid form. The next step is to identify some possible molecules – called 'biomarkers' – which, according to biologists, would be diagnostics of the presence of life. According to thermochemical and photochemical atmospheric models, the presence of a large fraction of oxygen (as on Earth) could not be explained by abiotic processes. Only a small fraction (less than 1 per cent as observed on Mars, even less on Venus) can result from the photochemistry of $CO_2$ and/or $H_2O$. So Tobias Owen suggested in 1980 that the 0.760 μm band of $O_2$ should be searched for in exoplanetary atmospheres. However, this band is weak and the visible range is not ideal for detection, because of the very low signal contrast between the planet and the star (we have seen that it is about $10^{-7}$ in the case of Jupiter and the Sun). A better diagnostic appears to be ozone, the dissociation product of oxygen, which exhibits a very strong spectral signature in the infrared range, at 9.6 μm (Figure 5.18).

In the future, spectroscopic observations of exoplanets during secondary transits should offer a good opportunity to search for this signature. We have to keep in mind, however, that the presence of large amounts of oxygen implies the development of an elaborate form of life, whereas living organisms may well exist without exhibiting a strong oxygen signature. Before oxygen appeared in the Earth's atmosphere, other processes took place, leading to the formation of methanogens, sulfatogens, and other products. In the case of Earth, most of the methane present in the atmosphere (about 1 part per million below 30 km) comes from the decomposition of living organisms. But the presence of methane can also be explained by abiotic processes, as we witness in particular on Titan and the giant planets, so it cannot be considered as a biomarker by itself.

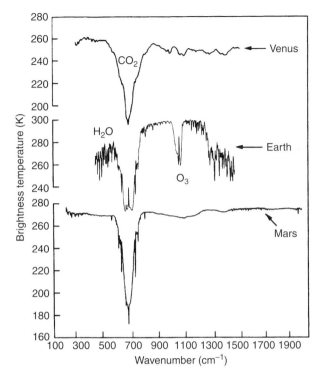

FIGURE 5.18 The infrared spectra of Venus (top), the Earth (middle) and Mars (bottom) in the infrared range. The presence of ozone at about 9.6 μm (wavenumber 1040 cm$^{-1}$) is the signature of life on Earth. In the future, this signature will be searched for in the infrared spectra of exoplanets. (Redrawn from Hanel *et al.*, 1990.)

Astronomers and biochemists have developed models to try to determine what combination of molecules could be considered a reliable diagnostic for the presence of life on an exoplanet. They found that the combination of $CH_4$ and $O_3$ would be a convincing indicator for life. The 6–16 μm range, well adapted for exoplanet spectroscopy in secondary transits, exhibits strong spectral lines from $CH_4$ at 7.7 μm, $O_3$ at 9.6 μm, but also $H_2O$ at 6.2 μm, $NH_3$ at 10.5 μm and $CO_2$ at 15 μm.

Another diagnostic has been proposed by other scientists to detect the presence of vegetation directly on rocky exoplanets. The idea is to measure the visible slope of the exoplanet's spectrum where

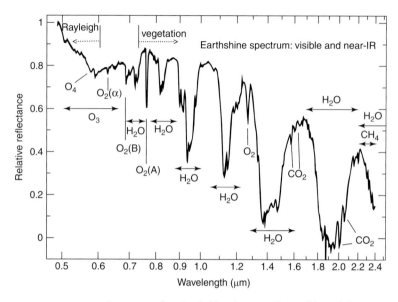

FIGURE 5.19 Spectrum of an Earth-like planet in the visible and the near-infrared ranges, where the red vegetation edge shows up around 0.7 micrometres. (Redrawn from Turnbull *et al.*, 2006.)

chlorophyll exhibits a sharp edge around 0.7 μm, with a strong absorption below this wavelength: the transmission increases from 5 per cent at 0.68 μm to 50 per cent at 0.73 μm. As a demonstration of this method, a search was made for this 'red vegetation edge' (RVE) in the Earthshine of the Moon during the Galileo flyby of the Earth (Figure 5.19). The experiment, however, is difficult: the detection strongly suffers from the presence of water clouds all over the planet (as would be the case, presumably, for any ocean-bearing exoplanet).

Exoplanet exploration is moving at a fast pace. New detections are reported everyday, but beyond that, we are currently in a position to attempt to categorize and to characterize them. Models of their composition and dynamics are being developed, and all of this is extremely exciting for planetary scientists and astronomers, for whom it is as if we are rediscovering the Solar System. In the next chapter we look at advances we may expect in the near and long-term future.

# 6   Extraterrestrial habitable sites in the future

## 6.1   FUTURE EXPLORATION OF POSSIBLE HABITATS

### 6.1.1   *Exploring the Solar System remotely and* in situ

Our exploration of Solar System habitats will continue with our usual astronomical means: ground-based telescopes, space observatories and *in situ* missions. Following the present generation of 10-metre class telescopes, astronomers are now working on the next step: a 30–40-m telescope. Three projects are currently being studied, two in the United States and one in Europe. On the American side, the GMT (Giant Magellan Telescope), made of seven 8-m telescopes, to be installed at Las Campanas in Chile, will reach an equivalent diameter of 21 m; the Thirty Meter Telescope (TMT), to be installed at Mauna Kea Observatory in Hawaii, will consist of a primary mirror composed of 492 hexagonal segments of 1.45 m diameter. Finally, the European ELT (E-ELT; Figure 6.1), to be installed at Cerro Armazones in Chile, will reach a diameter of 39 m by using about 800 hexagonal elements of 1.45 m diameter each. The first light of the E-ELT is planned for 2021. The E-ELT will be of special interest for the spectroscopy of transiting exoplanets, but will also be very useful for Solar System exploration, in particular the study of comets and trans-Neptunian objects.

As a follow-up to the HST, the New Generation Space Telescope, now renamed the James Webb Space Telescope (JWST; Figure 6.2) is built by NASA in partnership with ESA. This 6-m diameter telescope will be dedicated to infrared astronomical observations, from 0.6 to 28 μm. The spacecraft will be located at a special orbital position known as the L2 Lagrangian point of the Sun–Earth system, beyond the Earth on the Sun–Earth axis, a stable position much favoured by astronomical spacecraft. Its main scientific objectives are cosmology and

FIGURE 6.1 The E-ELT (Extremely Large Telescope) of about 40 m equivalent diameter, under development at ESO (artist view). It will be located at Cerro Armazones, at 20 km from Cerro Paranal, the site of the VLT, in Chile. Its first light is scheduled for the beginning of the 2020s. (Image credit: Swinburne Astronomy Productions/ESO.)

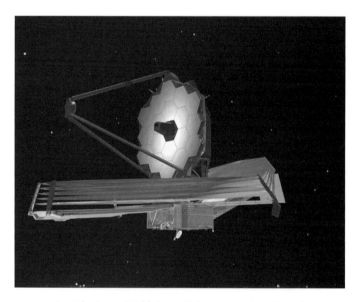

FIGURE 6.2 The James Webb Space Telescope project. For colour version, see plates section. (Image courtesy of NASA.)

exoplanets, but will also be a prime tool for Solar System exploration, especially for small bodies in the outer Solar System. It is expected to be launched in 2018 on an Ariane 5 rocket.

For planetary exploration, new projects are under development. The ExoMars mission, in cooperation between ESA and Russia, is made of two elements: the Trace Gas Orbiter (TGO) will be launched in 2016 for an in-depth analysis of the trace atmospheric constituents of Mars, and a descent module will land on the surface; two years later, a rover will carry a scientific payload devoted to astrobiological studies. In particular, the rover will carry a drill able to extract samples at a depth of about a metre, and will analyse them in a complex laboratory including chromatographs and mass spectrometers. Following the success of the Mars Science Laboratory (Curiosity) mission, NASA has plans for a new rover, even more sophisticated, to be launched in 2020. This rover would be used for identifying and possibly selecting Martian samples which would be brought back to Earth in a later step for ground-based laboratory analysis: this is the old dream of the Mars Sample Return mission, long desired by planetologists, but so complex to implement that its achievement has been delayed again and again. Concerning the inner Solar System, another ESA mission, in partnership with the Japanese agency JAXA, will be launched in 2015 to reach Mercury in 2022 and monitor the planet with two orbiters for an in-depth analysis of its surface, its magnetic field and its interaction with the solar wind.

The outer Solar System will be also a privileged target for space exploration. First, the Rosetta mission will approach comet Churyumov–Gerasimenko in 2014 and will deposit a lander at its surface in November 2014 – an impressive technological challenge. The New Horizons mission, launched by NASA in 2006, will approach Pluto and Charon in 2015, to be later redirected toward another Kuiper Belt object. The JUNO mission, launched in 2011, will approach Jupiter in 2016; its main objective is not astrobiology, but an analysis of the elemental abundances in its interior, to better constrain its formation scenario. Further into the future, the JUICE mission, recently selected by ESA, will be launched in 2022 and will approach the Jupiter system in 2030, with special emphasis on the exploration of the Galilean icy satellites (see Subsection 4.2.3). More projects exist, not yet approved, to explore the Saturn system with special attention to Titan and Enceladus.

### 6.1.2   Exploring exoplanets from the habitability point of view

Studying extrasolar worlds remotely and characterizing the newly found planetary systems from space and from the ground is a very exciting prospect for astrophysicists. Considerable resources are being devoted both to their detection and to their characterization by means of large telescopes and future space missions.

The velocimetry method (see Box 5.1), initially limited to velocity differences of more than $1 \text{ m s}^{-1}$, is expected soon to reach velocity limits of $0.1$–$0.3 \text{ m s}^{-1}$, which means that it should be possible to detect Earth-mass exoplanets. Dozens of programmes are going on around the world, using 2-m to 10-m telescopes. The most ambitious future projects will include the ESPRESSO at the VLT, and later the CODEX and MIRES instruments, respectively at the 42-m E-ELT and the 30-m TMT, by the end of the 2010s.

Direct imaging can now be attempted from the ground, provided the planet/star flux contrast is optimized (by the choice of an M-dwarf and a giant exoplanet distant from its star). As a follow-up of the detection of 2M1207b obtained with the NACO system at the VLT, by a combination of adaptive optics and coronagraphy, the SPHERE instrument, expected to be in operation by the end of 2013, should improve the performance of NACO by an order of magnitude. In the United States, the Gemini Planet Imager is expected to have comparable performance. Further into the future, the EPICS instrument, an improved version of SPHERE, is expected to be mounted at the E-ELT in the 2020s.

Astrometry from the ground has long been beyond the present instrumental capabilities, but astronomers are now developing new programmes on the basis of interferometers equipped with adaptive optics, with the aim of reaching limits of 10–100 milliarcsecs (mas).

A tentative but unconfirmed detection was reported by the STEPS (Stellar Planet Survey) at Mt Palomar. Among other programmes, the ESPRI project (Exoplanet Search with PRIMA) has started a 5-year survey using the PRIMA (Phase-References Imaging and Microarcsecond Astrometry) at the VLT. Attempts have also been made to use space means for astrometry, such as the Hipparcos New Astrometric catalogue, which gave marginal results. The HST Fine Guidance Sensor was used to constrain the geometry of a few planetary systems. But the future of exoplanetary exploration by astrometry definitely relies on the Gaia space mission, a European project presently under development which should be launched by the end of 2013. Gaia should be able to discover several thousand giant exoplanets at distances of 3–4 AU from their stars out to 200 parsecs, and is thus expected to achieve a major milestone in the history of exoplanet exploration.

After the successes of CoRoT and Kepler, new space projects are under development to detect and characterize exoplanets by transit. As we have seen, bioastronomers are particularly interested in those extrasolar planets that are located in the 'habitable zone' of their host star, where liquid water, acting as a solvent, enhances the contacts among molecules, thus leading to longer prebiotic chains. CHEOPS (on the European side) and TESS (on the US side) are two selected space missions devoted to a more efficient search for exoplanets, observing brighter stars and more extensive categories of target. In the future, spectroscopic observations from space will be privileged, in order to take full advantage of the mid- and far-infrared range. Scheduled for launch in 2018, JWST (James Webb Space Telescope), the HST successor, will be equipped by several imaging spectrometers covering the whole infrared spectral range from 1 to 30 μm, and should, among other things, permit the identification of more of those exoplanets located within the habitable zones of their stars. If one considers the powerful telescopes planned for even further into the future on space observatories, there is no doubt that we shall one day be able to make high-resolution observations of the atmospheres of these planets and

FIGURE 6.3 Potential design for ATLAST, a 16-m, segmented-mirror future telescope. For colour version, see plates section. (Image by Bill Oregerle (NASA/GSFC and Marc Postman (STScI), courtesy of NASA/STScI.)

determine their characteristics while at the same time searching for biosignatures. One such project is NASA's Advanced Technology Large-Aperture Space Telescope (ATLAST; Figure 6.3), a large optical/ultraviolet space telescope for the space observatory UVOIR with a segemented mirror of up to 16 m in diameter and considerably higher angular resolution and sensitivity than even JWST.

Other space missions, currently in the assessment phase, have also been proposed to characterize exoplanetary atmospheres and search for the presence of oxygen and ozone, as well as other species, using primary and secondary transit spectroscopy. An example is the EChO (Exoplanet Characterisation Observatory) mission, submitted to ESA as a candidate of the Cosmic Vision programme in the third call for proposals (M3) EChO is designed to observe the whole exoplanetary spectrum from the visible up to the mid-infrared at 16 μm. The expansion to longer wavelengths is needed to access not only hot Jupiters but also temperate exoplanets (the more promising targets for exobiology) which show a more favourable planet/star flux contrast.

## 6.2 PROTECTING POSSIBLE HABITATS

We have shown so far in this book that possible habitable worlds exist in our Solar System and beyond. We have also discussed where they may be found and how to study them in the future by means of large telescopes but also by *in situ* missions. But are we sure to be measuring the local environment once we get there? Before we consider making changes to such planets in order to transform them so that they can host humanity (see next sections), we may want to start by paying the courtesy of protecting these distant words from contamination while we investigate them. Space agencies all over the world have put together an important programme called 'Planetary Protection' with which scientists hope to preserve any possible lifeforms that do exist outside our planet by establishing procedures to ensure a lack of contamination.

Planetary protection works to protect Solar System bodies from contamination by terrestrial organisms brought in by space missions, and in turn to avoid contaminating the Earth with lifeforms that might be returned from elsewhere. Planetary protection is not to be taken lightly. If we want to be able to explore the Solar System without affecting and destroying the natural states of its planets, moons and comets, and search meaningfully for living organisms, we need to be sure that we are not interfering with their environments; and vice versa, we have good reason to ensure that we do not bring back contaminants harmful to the Earth's biosphere.

### 6.2.1 International treaties and organizations with relevance to planetary protection

As long ago as 1967, the United Nations 'Treaty on Principles Governing the Activities of States in the Exploration and Use of Outer Space, Including the Moon and Other Bodies' stated that all countries that had signed the treaty 'shall pursue studies of outer space, including the moon and other celestial bodies, and conduct exploration of them so as to avoid their harmful contamination'. An important player in this aspect on an international level, setting the practicalities for planetary protection, is the Committee on Space Research (COSPAR) which is part of the International Council of Science (ICSU) and has a dedicated panel assigned the task of making recommendations on the policy that should be adopted for planetary protection on a case-by-case level. ICSU collaborates directly with the United Nations in the planetary protection area. National agencies such as NASA, ESA, CNES, JAXA and CSA have established their own planetary protection policies that they apply to all their space missions, generally in accordance with the COSPAR guidelines. The agencies take very seriously into consideration all information that exists and the incoming updates relative to habitable conditions on planetary objects, and accordingly issue requirements for how a mission's design should take into account the problem of possible contamination. This is of particular interest for missions involving sample return or for those targeting bodies considered to be possible habitats.

### 6.2.2 Requirements for protecting life on other bodies

NASA's Office of Planetary Protection stipulates in its *Protecting Life on Other Bodies* guidelines (http://planetaryprotection.nasa.gov/), when discussing the possible categorization of the Solar System objects, that:

> Each mission is categorized according to the type of encounter it will have (e.g. flyby, orbiter or lander) and to the nature of its destination (e.g. a planet, moon, comet, or asteroid). If the target body has the potential to provide clues about life or prebiotic chemical evolution, a spacecraft going there must meet a higher level of cleanliness, and some operating restrictions will be imposed. Spacecraft going to target bodies with the potential to support Earth life must undergo careful design planning, stringent cleaning and sterilization, and submit to greater operating restrictions.

The following tables from the COSPAR documents on Planetary Protection Policy (COSPAR 2008) and NASA's Planetary Protection Programme indicate the current evaluation of such risks.

Table 6.1.

| Types of planetary bodies | Mission type[1] | Mission category[2] |
| --- | --- | --- |
| Bodies 'not of direct interest for understanding the process of chemical evolution or the origin of life'. | Any | I |
| Bodies of 'significant interest relative to the process of chemical evolution and the origin of life, but where there is only a remote chance that contamination carried by a spacecraft could compromise future investigations'. | Any | II & II* |
| Bodies of significant interest to the process of 'chemical evolution and/or the origin of life', and where 'scientific opinion provides a significant chance that contamination could compromise future investigations'. | Flyby, Orbiter Lander, probe | III <br> IV[3] |
| Earth-return missions from bodies 'deemed by scientific opinion to have no indigenous lifeforms'. | Unrestricted Earth-return | V (unrestricted) |
| Earth-return missions from bodies deemed by scientific opinion to be of significant interest to the process of chemical evolution and/or the origin of life. | Restricted Earth-return | V (restricted) |

[1] If gravity assist is utilized during a flyby, constraints for the planetary body with the highest degree of protection may be required.

[2] For missions that target or encounter multiple planets, more than one PP category may be specified.

[3] Category IV missions for Mars are subdivided into IVa, IVb, and IVc.

The best way to avoid contamination is always to avoid encountering a planetary object. Careful mission design and planning can help with this requirement. But sometimes we need to land or to crash on a target. For example, at the end of an orbiter mission, the spacecraft may be sent into the object, as the simplest solution or for scientific reasons. In that case, the spacecraft could be placed into a long-term orbit so that radiation and other elements of the local space environment may protect itself by eliminating any inadvertently transported Earth microbes.

In all cases, missions must meet stringent cleanliness and sterilization requirements, and these are expected to hold throughout the mission.

For probes such as landers and rovers intended to land on target bodies that present biological or habitable potential, the sterilization procedure may be more stringent. They can be designed so that only some parts are exposed to the surface of the target, especially if they carry life-detection experiments.

Table 6.2. *Specific planetary targets for all mission categories*

| Planetary targets/locations | Mission type | Mission category |
|---|---|---|
| Undifferentiated, metamorphosed asteroids; Io; others TBD. | Flyby, orbiter, lander | I |
| Venus; Earth's Moon; comets; non-Category I asteroids; Jupiter; Jovian satellites (except Io and Europa); Saturn; Saturnian satellites (except Titan and Enceladus); Uranus; Uranian satellites; Neptune; Neptunian satellites (except Triton); Kuiper Belt objects (<1/2 the size of Pluto); others TBD. | Flyby, orbiter, lander | II |
| Icy satellites, where there is a remote potential for contamination of the liquid-water environments, such as | Flyby, orbiter, lander | II* |

Table 6.2. (cont.)

| Planetary targets/locations | Mission type | Mission category |
|---|---|---|
| Ganymede (Jupiter); Titan (Saturn); Triton, Pluto and Charon (Neptune); others TBD. | | |
| Mars; Europa; Enceladus; others TBD (Categories IVa–c are for Mars). | Flyby, orbiter | III |
| | Lander, probe | IV (a–c) |
| Venus, Moon; others TBD: 'unrestricted Earth return' (unrestricted) | Unrestricted Earth return | V |
| Mars; Europa; Enceladus; others TBD: 'restricted Earth return' | Restricted Earth return | V (restricted) |

From these tables, it is clear that in the case of Europa, for instance, if its possible underground ocean is home to a form of life, the risk of contamination raises many ethical and ecological questions. Currently, these issues are already being discussed for future exploration missions to the satellite (the so-called JUICE orbiter to be sent by ESA and any future lander). Such a mission would be categorized III–V, similar to a mission to Mars.

## 6.3   FATE OF THE SOLAR SYSTEM AND EVOLUTION OF THE HABITABILITY ZONE

Our Sun is currently 4.6 billion years old, halfway through its main-sequence phase, fusing hydrogen into helium in its core, which will last a total of about 9 billion years. While in its main-sequence phase, the Sun will be losing mass and increasing in luminosity, but the only effect this will have is to disturb the orbits of the inner planets, causing chaotic motion, while it is likely that Mercury may come to a close encounter with Venus. When this stage is over, the Sun will enter the red giant branch (RGB) phase of the Hertzsprung–Russell diagram, expanding so much that its radius will engulf the inner planets, possibly including the Earth, although there are still uncertainties on what that final radius will be. In the next phase, the asymptotic giant branch (AGB) stage, the Sun will become very unstable, increasing its luminosity to several thousand times today's, before it blows off its outer layers and half of its mass in a planetary nebula while the temperature of the remaining core rises to over 100 000 K. Finally, the Sun will shrink to the size of the Earth, as a white dwarf.

There is no question but that these last dramatic phases of the Sun's evolution will have devastating effects on objects in the Solar System. Even though it is expected that the eight planets will remain in place, rotating around the dying Sun, we do not know if our Earth will survive the Sun's moods.

Some theories predict the loss of our home planet. The results from evolutionary theories are varied, and the Earth's fate is rather controversial, but Mercury and possibly Venus and Earth are expected to be swallowed by the expanding Sun during its red giant phase. Indeed, according to some models, the closest encounter of planet Earth with the photosphere of the Sun in its cool giant form would occur during the tip of the red giant branch phase. During this critical episode, considering the loss of orbital angular momentum suffered by the Earth from tidal interaction with the giant Sun, as well as dynamical drag in the lower chromosphere, simulations show that planet Earth might not be able to escape engulfment, despite the positive effect of solar mass-loss. In order to survive this destructive phase, any hypothetical planet would require a present-day minimum orbital radius of about 1.15 AU. So Venus, Earth and perhaps even Mars may no longer be viable habitats.

It is clear, then, that the habitable zone as we know it today will evolve and move outwards. We may then have to look to the outer planets and their moons for habitats. Indeed, it has been argued since 1997, by astrobiology experts such as Ralph Lorenz, Jonathan Lunine and Christopher McKay, that Saturn's satellite Titan, for instance, may then become habitable. Titan's massive atmosphere and complex surface, with all their significant astrobiological implications, have been discussed in Chapter 4. As these authors say in their paper

(Lorenz *et al.*, 1997), the response of Titan's surface and massive atmosphere to the change in solar spectrum and intensity as the Sun evolves into a red giant is small as the haze-laden atmosphere expands and blocks more sunlight. These studies indicate that, as the haze production drops owing to the increase of the ultraviolet flux from the reddening Sun, a window of several hundred Myr opens, roughly 6 Gyr from now, when oceans of liquid water–ammonia could form on the surface and react with the abundant organic compounds there. As Lorenz, Lunine and McKay say:

> ...the duration of such a window exceeds the time necessary for life to have begun on Earth. Similar environments, with 200 K water–ammonia oceans warmed by methane greenhouses under red stars, are an alternative to the 300 K water–carbon dioxide environments considered the classic 'habitable' planet.

In spite of all that, it may well be that according to a pessimistic scenario of the evolution of our Solar System, it will completely cease to exist when our Sun has cooled down to 5 K, and the gravitational attraction of passing stars will strip our star of its planets.

Humanity will then have to move, but it might be a good idea for our grandchildren to start making plans before that and to explore methods to occupy other planets, moons and asteroids of the outer Solar System well before the Sun leaves the main-sequence phase – all the more so if we humans continue our destructive influence on our planet. Indeed, Earth may be wracked by the consequences of over-population and climate change in 300 years or so.

We explore hereafter some of the housing options humanity might have elsewhere in the Solar System. But first, a word of caution...

## 6.4  HUMANS IN SPACE

We, the people of Earth, are always interested in searching for life outside our planet, but at the same time we should be (or become more and more) concerned about our own habitat, the Earth's environment. As we saw in Section 6.3, the Sun's evolution will cause the conditions in our Solar System to become unfavourable or even unbearable for human beings at distances closer than Mars, and later on even further away. Humanity (or whatever thinking beings exist in four billion years) might then have to consider ways of surviving by moving to the outer realms of the Solar System or even beyond. In order to do so, space colonization could appear a necessary step.

The situation is not that simple, however. The question of space colonization is actually raised by motivations other than the survival of human beings in a few billion years. They are multiple, and deal with many different aspects: the dream of humanity, the attraction of human-kind to discovering new worlds, the prestige of a nation, its economical and political power. Among all these reasons, science is far from being a major one; this is why the scientific community is deeply divided about space colonization. Many scientists consider that, given the amazing successes of robotic exploration, both for planetary and astro-nomical missions, manned exploration is a dubious choice, particu-larly in view of its overwhelming cost.

In what follows, we review the arguments for and against manned exploration, on the basis of the history of the past 50 years of space exploration, looking at some medical, engineering and survival arguments in favour, but also at the low scientific return, huge invest-ment and the changing evolution of the Earth's climate which requires other priorities. However, no matter what scientists prefer, there is a good chance that manned exploration will take place at some pace, because of the technological challenges associated with it and its huge political implications. So, in the next sections, we will explore some possibilities that not only have been imagined by science fiction writers but are the result of serious studies by space agencies and institutions all over the world. We will review current and future options for human beings to live and survive in space and/or on extra-terrestrial worlds, still keeping in mind that we are often at the limit of science fiction, or even fully within its realms.

### 6.4.1 Manned exploration: should we privilege it?

The first landing of a man on the Moon by Apollo 11, on 21 July 1969, marked the triumph of manned exploration. This event concluded the sustained race between the USA and the Soviet Union, initiated in 1957 by the successful launch of Sputnik. In the middle of the cold war, the success of Apollo was tremendous both from the technological and from the political point of view. The Apollo programme was nevertheless prematurely terminated three years later with the final Apollo 17 mission. The US politicians soon realized that, after that prestigious success, there was not much to be gained. Science, indeed, was only a minor objective for the Apollo programme, in spite of what may have been said to justify its cost. As an illustration, the Soviet Lunakhod robots managed to return lunar samples to Earth without men on board.

What has happened since then and during the past 40 years? Because of its huge costs, lunar exploration was stopped. NASA decided to develop the space shuttle, supposed to reduce these costs enormously. The programme was seriously hampered, however, by the two tragic losses of the shuttles Challenger on 28 January 1986, and Columbia on 1 February 2003. The Soviets, in the meantime, concentrated on manned flights in Earth orbit, with several generations of orbital stations; their greatest success was the MIR station, in orbit between 1971 and 2001. In 1984, the US President Ronald Reagan announced the development of a large international station, which would become the International Space Station (ISS; see following section). Again, the objective was mostly political: to maintain an American presence in space at the same time as the Soviet MIR station and work on a common international project in order to prevent future international conflicts among the largest countries. The European Space Agency joined the project in 1988 and Russia in 1993. The scientific utility of the ISS remains controversial, as will be discussed in the following section. Although some scientific and engineering studies have been conducted to evaluate the possibility of constructing 'space cities' (see Subsection 6.4.3) in order to compensate for some of the issues faced by the ever-increasing Earth population, these considerations remain far-fetched for the moment.

But space exploration does not stop at the Earth's orbit. Many people (including some scientists, but certainly not all) are strong advocates of sending people to Mars, and several projects to that end have been studied in some detail (see Section 6.6). Again, the cost of the travel alone appears prohibitive. Some NASA studies state that the 'space train' from Earth to Mars would weigh several hundreds of tonnes, 10 times more than a manned mission to the Moon. Landing on Mars will be much more complicated than on the Moon, owing to the presence of an atmosphere which acts as a thermal shield. And of course, establishing a colony on Mars, as for the Moon, requires building a base, reconstructing a viable confined atmosphere, and bringing in all the necessary resources, including electric power and, obviously, water...

As for leaving behind the Solar System to develop extrasolar colonies, we are now entering the realms of science fiction: so far, we do not know of any means of propulsion that could transport human beings all the way to the nearest star within a human lifetime. The nearest stars are several light-years away. Even supposing that we could travel at a speed as high as a tenth of the speed of light (which of course we are incapable of doing today), the trip to a nearby star would last almost the whole of an unfortunate astronaut's life.

In conclusion, there are very sound reasons not to favour manned missions. We could also evoke the problem of climate warming, which requires that a maximum of resources should be devoted to saving Earth and our living environment. But in spite of these arguments, as mentioned above, it is likely that the history of manned exploration will not stop yet. So, below, we describe the various initiatives that have been developed or are under study for the near and distant future.

### 6.4.2 International Space Station: taking humans into space

The International Space Station (ISS; Figures 6.4, 6.5) is an unprecedented achievement in global human efforts to design, construct, operate and exploit a research platform in space. Its assembly began in November 1998. International space agencies, most prominently

FIGURE 6.4 The International Space Station is an international research platform in space, launched on 31 October 2000, and docked 2 November. In the past 10 years of continuous human occupation, the space station has been visited by more than 200 individuals over the course of more than 57 000 orbits around the Earth. (Image courtesy of NASA.)

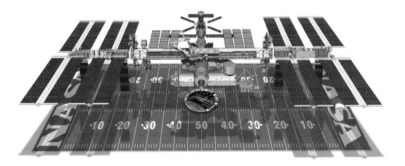

FIGURE 6.5 The space station, including its large solar arrays, spans 120 m in length and weighs about 420 tonnes, not including visiting vehicles. The complex now has a habitable volume of about 400 cubic metres. (Image courtesy of NASA.)

including the United States, Europe, Russia, Canada and Japan, work together to provide and operate the ISS elements. Each agency is responsible for the functioning and the management of the hardware it provides, and therefore the ISS is a complicated programme from the political point of view. By July 2012, and following the launch of the first module, Russia's Zarya, on 20 November 1998 the ISS had received 125 visits mostly by Russian vehicles and space shuttles carrying more than 200 individuals. The International Space Station Programme is therefore a common venue for international astronauts and scientists from around the world, docking multi-national vehicles and facilitating exchanges on operations, training, engineering, communications networks and development facilities. The ISS is an example of collaboration and cooperation on a large scale.

In the past decade of its existence, the space station has performed more than 57 000 orbits around the Earth covering more than 2.4 billion kilometres in distance (about ten round trips from the Earth to the Sun). The elements that compose the space station were assembled in orbit, and this required some 160 spacewalks that took about 1000 hours. The final space shuttle mission in this assembly phase took place in July 2011, when the space shuttle Atlantis delivered around 4.5 tonnes of supplies for the Raffaello logistics module and the AMS-02 instrument. One more pressurised Russian module remains to be attached to the station in the future.

The space station, including its large solar arrays, is about 120 metres in length and weighs about 420 tonnes, not including visiting vehicles. The power generation is about 84 kilowatts. The complex currently affords a habitable volume of about 400 cubic metres, a total pressurized volume of 920 cubic metres, and boasts two bathrooms, a gymnasium and a 360-degree bay window, enough room for its crew of six persons and a vast array of scientific experiments.

In particular, the 2005 and 2010 NASA Authorization Acts designated the US segment of the space station as a national laboratory using, among other things, its unique capabilities as a permanent microgravity platform exposed to the space environment. Some of the research goals for the space station have to do with physical sciences, biology, human health, and technology testing for future exploration and for earth and space sciences in general.

The European Space Agency is responsible for two key station elements: the European Columbus laboratory, specializing in research on fluids, different materials and life sciences, and the Automated Transfer Vehicles (ATV), a supply ship put into orbit on an Ariane-5 launcher.

The European Programme for Life and Physical Sciences (ELIPS) which started in 2001 is a platform allowing Europe to become an efficient user of the ISS, leading to many research advances in current and future spacecraft activities and in scientific disciplines such as health research, innovative materials and technologies, plasma physics or exobiology. By using the European Columbus laboratory as well as other ISS and additional research platforms (ground-based, drop towers, parabolic flights and sounding rockets) the second phase of ELIPS will enhance this research and add new activities for future use.

The ISS participating countries share a cost of about 100 billion euros over a period of almost 30 years. Rather like an outpost on the Moon, some consider the ISS as a step bringing humans closer to colonising space, but others have reservations.

In view of the tremendous cost of the ISS project, many voices among the scientific community have questioned its utility. From a technological and political point of view, the ISS is undoubtedly a big success. But what is its scientific purpose or its end product? The main results come from the study of the human body reactions to microgravity in a confined environment. From a research and medical point of view, this work is certainly useful (see following section), although it seems that the question as to whether the human body is adapted for living in space environments has already been answered in the negative. Our body is the product of a slow evolution influenced highly by the evolution of environmental parameters (pressure, temperature, hygrometry...) but also from the beginning fundamentally by gravity. Although the studies conducted on the ISS have significant applications in the medical sciences beyond our adaptation to living in space, the conditions for such experiments might also have been created on Earth. Is it then reasonable to spend so much money to try to compensate these effects in space, or to conduct experiments when most of the scientific objectives could be achieved at a lower cost on our planet or with robotic exploration? The cost of the ISS is equivalent to 100 medium robotic missions or 30 large ones that would take us to the outskirts of the Solar System. In addition, the success of miniaturization techniques over the past decades have allowed us to reproduce the most complex robotic functions in ever reduced volumes, with accordingly decreasing cost. The human body, by definition, cannot be miniaturized.

### 6.4.3   Space cities

The idea of space cities and in particular of orbiting cities has been carefully considered not only by science fiction writers but also by astroscientists, who viewed it as the first logical step in colonizing the skies, following NASA's 'Space Colonization Basics'. An orbital space settlement is a giant spacecraft, kilometres across, which can host several thousands of human beings, animals and plants. Orbital settlements will travel endlessly through space while the people inside go about their activities and live 'normal' lives. The settlements would rotate so that you would feel something very close to Earth's normal gravity at the hull, and remain at close distances from the home planet to aid transport of supplies or management of advanced repairs. Such enormous urban structures were seen as representing a possible solution to the Earth's problems of environmental destruction, overpopulation and limited resources.

So it was no wonder that at some point the idea became the subject of serious brainstorming encouraged by man's landing on the Moon in 1969. Beginning the next year, NASA financed an ambitious project to place an entire human community into space. In spite of NASA's efforts and prediction though, and although the International Space Station has partly fulfilled its purpose, humankind still lacks the long-term viable environment where it can have a chance to survive the sad fate of our overpopulated planet in any foreseeable future. But the projects and designs of space cities were valuable in challenging our imagination, and much thought went into these concepts.

To be useful to large populations, a space village or city would need a robust urban design to prevent chaos and human losses. Among other indispensable features of the initial concept are of course houses and businesses, but also public services such as hospitals, schools and recreation facilities, a transportation system and even some agriculture and industry. In such a closed environment one would have to ensure recycling of oxygen, water and waste materials, and protection from solar radiation.

During this conceptual project, scientists at the Ames Research Center and in other institutes (Stanford University in particular) imagined several possibilities for massive, orbiting cities that were first shown to be theoretically feasible by Princeton physicist Gerard O'Neill in the 1970s (Figures 6.6, 6.7). Designed to carry as many as 10 000 people, these remote abodes had to supply humans with all the

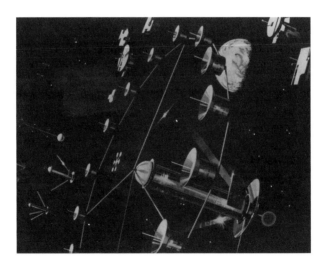

FIGURE 6.6 Space cities populating the near-Earth space. The concepts for these colonies were proposed by Princeton physicist Gerard O'Neill with help from researchers at Stanford and NASA's Ames Research Center, during an initiative for imagining living in space. (Image courtesy of NASA/AMES.)

(a)

(b)

FIGURE 6.7 (a) A view of O'Neill's space city (Cylinder) concept. The round rooms clustered tightly in a circle at the tip of each cylinder would be dedicated to agriculture, with adaptable humidity and temperature conditions. The designs shown here and in other conceptual images indicate that the size would be quite big (6 km or so). The sealed city also could afford to enjoy its own weather. The artist has imagined vast windows along the sides of the tubes which would allow for amazing views of the home planet, Earth, and of far-away regions of the Galaxy at the same time. (Image courtesy of NASA.) (b) A second type of orbiting city, the Stanford Torus. This doughnut-shaped colony was designed to be much smaller (by factors of 4) than the O'Neill concept, but still expansive enough to promote the growth of a significant space society. The lid-like item to the right is a gigantic mirror that focuses sunlight onto the colony, helping them survive in space. Such artistic views, many the work of Donald Davis, have been based on real engineering concepts. (Image courtesy of NASA Ames Research Center and Stanford University.)

necessary commodities for long stays. To simulate everyday conditions on the Earth, some sort of gravity would have to be created through the rotation of the edifice. Thus people could walk around feeling what would almost seem like gravity. The O'Neill Cylinder, imagined by Princeton physicist Gerard O'Neill, consisted of two cylinders flying in parallel formation at a pre-defined location between the Earth and the Moon – the so-called 'Lagrangian libration point' – at a spot where exchanges with the mother planet would be feasible as needed.

Another concept was the Stanford Torus (Figure 6.7b). These ultra-exclusive toroidal cities would contain luxurious residential areas and shopping centres, and inhabitants might use the riches of space to finance their abodes.

An third variety of orbiting city that NASA conjured up in the summer of 1975 during a workshop on such ideas was the 'Bernal Sphere' (Figure 6.8a), where large numbers of humans co-exist in the interior surface of a large sphere, about a kilometre in circumference. The entire sphere would rotate about twice per minute, producing a centrifugal force used as a substitute for gravity, which would be similar to what we have on Earth at the equator, diminishing gradually to zero at the poles. Natural sunshine would be brought in through external mirrors. Inhabitants could have the 'weather' they preferred, without worrying about its effect upon the crops: agriculture would be conducted in neighbouring buildings, outside the spherical portion of the habitat.

Should one wonder about the access to space, the space elevator (Figure 6.8b), a stationary tether rotating with the Earth, held up by a weight at its end above the geostationary orbit, is a concept that even today attracts considerable interest (the Annual Space Elevator conference was held again in August 2012 in Seattle). Upward centrifugal force created by the Earth's rotation would cause the cable to remain stretched taut, compensating for the gravitational pull. Once above the geostationary level, anything climbing along the tether would feel a force in the direction away from the Earth as the centrifugal force

(a)

(b)

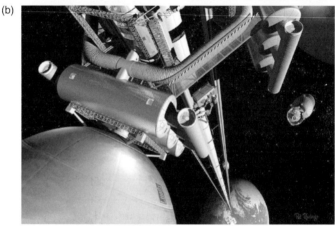

FIGURE 6.8 (a) The Bernal Sphere is a concept for hosting large numbers of humans upon the interior surface of a large rotating sphere, about 1 km in circumference, which would use centrifugal force as a substitute for gravity. For colour version, see plates section. (Image courtesy of NASA.) (b) The space elevator concept consists of a cable or tether fixed to the Earth's equator, reaching into space to facilitate space travel. For colour version, see plates section. (Image by Pat Rawling, courtesy of NASA/MSFC.)

overpowers gravity. According to its supporters, with such space elevators, travel to space might become commonplace. Indeed, in the absence of the pull of gravity, need for fuel on board and rockets to launch, it may be a simpler and cheaper means to transport a lot of heavy payload. The technical details, of course, would have to be worked out and may be more complex than thought, but the concept is also applicable to other planetary objects. For instance, Kevlar was considered for the cable in some of the original designs as being both robust and light; but at places in the Solar System with weaker gravity than Earth's (such as the Moon or Mars), the tether would be easier to put in place with less stringent constraints on its material.

More ideas and artistic depictions can be found in John Metcalfe's 2012 paper in the *Atlantic Cities* (http://www.theatlanticcities.com/).

Coming back to the space cities, the low-gravity conditions that they could obtain at their centres was considered as a major selling point for all those sensitive to the beneficial effects for the body of weightlessness. But other considerations had to be taken into account, such as the fact that large diverse groups would need to live in confined spaces. Note, however, that this is something that we shall also have to learn to do on a future overpopulated Earth (more than seven billion people today and growing). Space settlements may, when conditions become dire, offer an alternative to hunger or war – the ability to live in fairly homogeneous groups, if we can choose our space communities.

Orbital space settlements are discussed, among other matters, on the National Space Society website: http://www.nss.org/settlement/space/. Although studies of the concepts were eagerly launched and settlements viewed possible as early as the second half of the past century, they do not seem so appealing today. Perhaps the local threat has not quite hit home yet, but one can safely bet that if human colonization becomes a reality some of these ideas will constitute a strong heritage.

For human colonization, other factors come into play when one considers long trips through space. Human spaceflight would then have many physiological and psychological issues to deal with, such as radiation and long travel duration. Space life sciences and physical science experiments and investigations are essential in preparation for long-duration flights. An understanding of the biological consequences of microgravity and radiation exposure will be significant to humans living in space. For this, it is necessary to bring together scientists, engineers and experts from different fields to discuss topics of inter-disciplinary character: protection from radiation, biofluids under microgravity, habitats and life-support systems, exploration of space and planetary resources, biochemical analysis and more. But the topic is something we may not be able to avoid.

## 6.5 TRANSFORMING ('TERRAFORMING') POSSIBLE HABITATS

Once living on Earth becomes truly impossible for humanity, we may need to look at moving to distant worlds that will have been trans-formed to support life. In particular, we will need to have created an uncontained planetary biosphere emulating all the functions of the biosphere of the Earth – one that would be fully habitable for human beings in a foreign environment.

We call this 'terraforming', which, according to the *New Shorter Oxford English Dictionary*, Vol. 2 (1993), means to transform through intense engineering a planet, environment, etc. into something resem-bling the Earth, especially as regards suitability for human life. This definition indicates that there is currently more in this concept than just the science fiction side. Another term, suggested by Haynes in 1990, was ecopoiesis, from the Greek οίκος, a home, and ποίηση, creation. It was then defined by Martyn Fogg in 1995 as 'the fabrication of an uncon-tained, anaerobic, biosphere on the surface of a sterile planet. As such, it can represent an end in itself or be the initial stage in a more lengthy process of terraforming.' Nowadays, visions like that are taken more seriously and discussed more extensively, as in the science fiction novel *2312* (Kim Stanley Robinson, 2012). As when you take up a new residence, ecopoiesis cannot just happen on its own – in the sense that living organisms cannot be extracted from their original places, set down in new, different and sometimes hostile ones and be expected to survive and evolve there. Even extremophiles need some help, and according to experts, some changes and adjustments are absolutely required, to help any organism, no matter how willing and robust, to adapt to its new home (as happened during the Precambrian period on Earth). This kind of 'planetary engineering' needed at the start and later on leading to ecopoiesis is the basis of most terraforming-related research.

Most of the current considerations and studies on terraforming by experts such as Martyn Fogg or Robert Haynes have been focused on Mars, the next planet from the Sun after Earth, which has always attracted attention from the public and scientists alike as a possible habitat. But we have come a long way, as we have tried to show in this book, in the past two centuries, and today potential habitats can be found in many corners of the Universe.

In the neighbourhood close to the Sun, we obviously look at the nearby planets: Mercury, Venus and Mars. We have shown that the inhospitable environment of Mercury can only become even more so with time. But what about Venus and Mars?

Early on, Venus attracted attention in this sense, although its massive $CO_2$ atmosphere and 450 °C surface temperature certainly do not appear favourable for colonization purposes. Carl Sagan, a famous planetologist well known for his outreach activities, among which were the *Cosmos* series and book, proposed back in 1961 the planetary engineering of Venus by seeding it with algae, which would convert the existing water, nitrogen and carbon dioxide into organic compounds, thus reducing the $CO_2$ in the atmosphere. Sagan argued that the resulting carbon-based organics would then be deposited on the extremely hot surface of Venus, and thus be burnt and transformed into 'graphite or some involatile form of carbon' on the planet's surface. As we saw in Chapter 3, the smaller and smaller amounts of $CO_2$ would cause the greenhouse effect to be reduced and the surface temperatures to drop to habitable values.

All of this sounded very promising at first, but later, more accurate discoveries of the conditions prevailing on Venus showed that the process was impossible in the presence of the highly concentrated sulfuric acid solution composing the thick clouds of Venus. Any atmospheric algae would be unable to survive for long in such a thick, hostile upper atmosphere. Furthermore, the high atmospheric pressure would allow only pure $O_2$, and not water, to remain while the planet's surface would quickly become entirely covered in graphite powder, and any carbon-based organics that could then have been formed would eventually lose their carbon through combustion to become $CO_2$ again, thus inhibiting the terraforming process. It would then seem that in the inner Solar System, terraforming would be best attempted on Mars. Attention to that planet for colonization purposes dates a long way back in humanity's history.

### 6.5.1   Runaway greenhouse scenarios for terraforming Mars

Even on Mars, to allow ecopoiesis to be applied in a Martian environment, some adjustments are unavoidable, as explained by Fogg. To begin with, the mean global surface temperature (218 K, equivalent to –55 °C or –67 °F) must be increased by ~60 K to reach common (chilly) values on Earth. An atmosphere must be created to supplement the small amount of gas existing on Mars, and the atmospheric composition must be altered to increase its oxygen and nitrogen abundances. Of course, one would need liquid water. And finally, the surface should not be subject to high values of ultraviolet and cosmic ray flux.

As we have seen in Chapter 3, Mars is scarred by many fluvial-like features which may mean that in its earlier history the planet afforded a much denser $CO_2$ atmosphere. Researchers have tried to characterize this initial environment, and the goal of many climate-driven models is to try to simulate it. There is quite some evidence today to suggest that Mars' early history included periods of warmer and wetter conditions. Observers, on the other hand, try to find indications of the presence of the $CO_2$ today on Mars in surface reservoirs, hoping that a small warming of the planet might cause $CO_2$ to be injected into the atmosphere from its surface stock, causing a greenhouse effect and heat transfer to the poles. The cycle of increase of surface temperatures and consequent further release of $CO_2$ could then start, and, as on Earth, this runaway greenhouse effect would increase

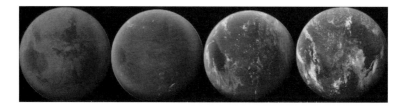

FIGURE 6.9 Images corresponding to an imaginary Mars transition towards an Earth-like planet. For colour version, see plates section. (Image by D. Ballard, courtesy of Wikimedia Commons.)

the pressure and temperature on the surface, making Mars more habitable (Figure 6.9).

The first Martian terraforming models were based on the faulty assumption that up to 1 bar-equivalent of $CO_2$ ice was stored in the polar caps, but as has been shown recently the Martian polar caps are composed principally of water-ice with perhaps just trace amounts of $CO_2$ or a $CO_2$ hydrate. However, it has been hypothesized that a non-negligible amount of $CO_2$ might exist adsorbed on mineral grains in the shallow subsurface of Mars. Whatever its location on Mars, warming the planet would allow the release of any trapped $CO_2$ ice which could then be used to create a runaway greenhouse effect, reversing the freeze-out of the Martian atmosphere, something known as 'the standard paradigm' of Martian ecopoiesis. As M. J. Fogg argues:

> an initial warming of the Martian surface by 5–20 K (depending on model parameters) increases the atmospheric pressure to a few tens of millibars at which point a runaway becomes established resulting in a stable end state of ~800 mbar and ~250 K. A 2 bar reservoir would runaway to give a mean surface temperature of ~273 K and a 3 bar reservoir, >280 K.

If $CO_2$ is not available, other gases such as CFCs (from the group of hydrofluorocarbons, which are made up of chlorine, fluorine, carbon and hydrogen) are long-lived, stable and non-toxic, and could trigger outgassing leading to a far stronger artificial greenhouse effect and producing up to 30 K warming of the planet. Studies of the conditions on Mars have shown, however, that CFCs on the red planet are far less stable and long-lived than on our own planet, owing to the absence of shielding of the surface by an ozone layer.

Other possible ways to increase the temperature of Mars have been imagined. One of these advocates the use of large orbiting mirrors to increase the amount of solar energy received, by reflecting light to the planet's surface. In the so-called 'solar sail' mirror systems, described by Robert Zubrin and Chris McKay in 1993, such a mirror 125 km wide and made of aluminium would have a mass of 200 000 tonnes and create a runaway greenhouse effect if stationed 214 000 km behind Mars, from where it could illuminate the south pole and provide it with an extra ~27 terawatts. This would cause the polar temperature to rise by ~5 K which, according to some models, should be sufficient for cap evaporation. Some ideas for space mirrors around 20 m across have already been tested in Earth orbit by the Russians, and that or other, larger alternatives could be envisaged for Mars.

In some of these runaway greenhouse scenarios for terraforming Mars, the prediction is that it could be transformed into a planet habitable for anaerobic life in roughly a century. Conditions similar to the harsh and cold Precambrian would then prevail which would be more hospitable than those on the present Mars, and further terraforming might follow ecopoiesis to oxygenate the atmosphere through photosynthesis. Suggestions of engineering alternatives to the runaway greenhouse exist. If $CO_2$ deposits do exist on the planet but are essentially bound up in carbonate minerals, then more extreme processes have been suggested for causing outgassing, including nuclear explosives or directed asteroid impacts. It is clear that such processes would be both destructive and difficult to control.

A more serious problem is the obvious lack of abundant water on Mars. Any known reserves of water are in the polar caps or at high latitudes in the subsurface, which makes them difficult to extract and inject into the biosphere, as heat conduction through the loose rock layer (the regolith) is very slow. However, some recent modelling of the

Martian hydrological cycle suggests that the mid-latitudes on Mars could hide aquifers in their interior that might be accessible by drilling. Even so, releasing the water would require much more energy than for $CO_2$, and all the above methods, including the most extreme, would have to come into play.

### 6.5.2 Terraforming in the outer Solar System: icy satellites and asteroids

The extreme actions cited above are not our usual idea of terraforming. Generally, when scientists speak of terraforming (Mars, for example), relatively small changes are imagined first that would create an environment suitable for life that would come from Earth. The outer Solar System only offers difficult possibilities, but still ones worth considering, as we have showed previously in this book. In particular, humans who have tried to live in microgravity on spaceships know that that represents a problem, and it is the reason that all space city projects are bent on creating pseudogravity conditions. But we do not know what would happen if we lived on low-gravity planets like our Moon, or an asteroid, or the giant planets' moons, where we have at our disposal land and water, but there could still be health impacts harmful to human beings. Perhaps temporarily returning to Earth (while possible) or going up to space stations would be a solution. Another issue with all those far-away but unprotected landscapes, especially Mars and all the big solid-body moons, except Titan, is the necessary shielding from harmful cosmic ray impacts.

The insides of the asteroids offer an abode that avoids both those problems, because interior artificial gravity can be created through spinning and the walls of the asteroid's interior offer the necessary radiation protection. Indeed, terraforming of asteroids has been envisaged in various scenarios where a hollow space created inside such a body could be used to host humans.

Furthermore, although the gas giants hold no real interest for preserving life as we know it on Earth, their satellites may in fact be viable for Earth colonization. Since the drop in temperature hinders possible habitats from existing anywhere further out than 10–15 AU (at least in any foreseeable future), we focus hereafter on the solid bodies around Jupiter and Saturn.

Europa, one of Jupiter's moons, is such a candidate, as we showed in Chapter 4. On its surface, gravity is 0.14g (where g is the Earth's gravity; on the Moon, it is 0.16g). Its very thin atmosphere is composed mainly of $O_2$ at a pressure of 0.1 micropascals (or $10^{-12}$ times the Earth's). Its mostly exposed surface is then subject to strong radiation from Jupiter, so severe that a human being would not survive for 10 minutes. This raises serious constraints on the exploration and colonization of Europa as we have seen in Section 5.2. The very extreme conditions prevailing on the planet's surface probably cannot be remedied by technology. But it is still possible that the depths of the planet harbour a salty, subsurface ocean, whether frozen or in liquid state.and perhaps even some form of life.

In Jupiter's neighbourhood, other moons such as Ganymede and Callisto might also be viable candidates to consider. Ganymede, as we have seen in Chapter 4, is the largest moon in the Solar System. It is believed that it might be a good candidate for space colonization, owing to its size, the presence of a magnetic field and the possibility that it may harbour liquid water under the surface. If all of this is true, Ganymede could be heated by the injection of greenhouse gases: sulfur hexafluoride has been suggested as a possibility. Even for Callisto, where terraforming seems close to impossible with current technology, it has been hypothesized that large amounts of nitrogen might be deposited, or created by converting the ammonia existing on its surface through nitrates, and nitrogen would act as a protective gas. But that is not enough, and in the case of Callisto large quantities of some kind of greenhouse gas would have to be transported to create a viable atmosphere. There is no lack of water-ice on Callisto, so that aspect of terraforming would be relatively easy. Once the injected greenhouse gases had begun warming up the satellite, the water could melt to form oceans and, through electrolysis, also produce the oxygen in the atmosphere. Added to the advantages of this large Jovian moon is that there is no need for protection from radiation, as the radiation levels there are low.

From all of the above, it can be seen that scientists and engineers do not lack ideas for creating habitats out of currently inhospitable environments. This is also true for the more distant satellites around Saturn. One of these in particular, Titan, is a very interesting prospect in terms of terraforming, as it has been quite often considered as an 'Earth-like world'. Arthur C. Clarke, in his novel *Imperial Earth*, imagined a planet similar to Titan, dominated by the Makenzie family. Could this dream become reality in the future?

Titan has many organic molecules in its atmosphere and on its surface, where large amounts of exposed water ice also exist, and is also thought to possess a subsurface liquid water–ammonia ocean which could potentially harbour life. This moon therefore has important astrobiological potential and its investigation is essential for exobiologists. The presence of an atmosphere already composed essentially of nitrogen and hosting an active organic chemistry argues in favour of easy terraforming. Indeed, although its surface gravity is about $0.14g$ (around a tenth of Earth's), its atmosphere is dense (1.5 times the terrestrial atmospheric pressure) and contains 98.4 per cent $N_2$, with about 2 per cent methane and other trace gases. Its surface is formed through the flow of liquid methane and ethane, the deposition of organics through the atmosphere (forming dunes) and the erosion of ice as winds blow from west to east, in the direction of the moon's rotation. Near the equator, mists of methane have been observed. Plate tectonics may still be active on Titan, giving rise to cryovolcanoes. These ice volcanoes, which would also inject methane into the atmosphere, could explain some of the observations made by the Cassini mission. Their existence and the presence of a liquid water ocean could be confirmed in the future if the signatures of water and ammonia are detected at their locations

But not everything is simple for Titan. The low gravity could cause serious medical consequences (such as decalcification, weakening of the immune system, pregnancy difficulties or paediatric problems). Inhabitants of Titan would probably adapt to their environment, but future generations might no longer really be capable of living on Earth. In addition, there is little or no free oxygen present in the atmosphere, very little sunlight that penetrates to the surface (1 per cent of that arriving on our planet), few if any silicate surface materials, and a surface, at 94 K (−178 °C), far too cold for liquid water to exist.

Indeed, the low temperatures are probably the biggest problem. As we have seen, for Mars relatively small changes to the atmosphere could create a partial greenhouse effect, because the temperatures on Mars are only slightly below those of Earth. Titan, on the other hand, is nowhere near the Earth's conditions, with freezing temperatures that mean it would need a vast source of energy to warm it up, and no magnetic field to protect it from solar wind, cosmic rays or radiation effects. To melt the surface ice water and evaporate it into the atmosphere, in the case of permanent settlements, would mean the necessity for considerable heating and therefore powerful energy generators. For terraforming then, the aim should be to increase greenhouse gas emissions significantly and to reduce the formation of the ice fog (which acts against greenhouse warming) thus heating up the surface and atmosphere.

Another problem is that Titan is immersed in Saturn's radiation field for a major portion of each orbit, and apparently does not possess a magnetic field to help to keep that radiation away from its atmosphere and surface. However, it does own a complex ionosphere and the haze offers some protection.

In conclusion, there are pros and cons in the case of Titan, as for those of the other icy moons, which need to be investigated in the future by the supporters of manned space exploration. In our current era, planetary engineering in the inner or outer Solar System is, as described by M. J. Fogg (1998),

> ...concerned mostly with defining the boundaries of the possible, rather than in charting some definite route into the future. The concept is still at the limit of fantasy, although confirmation of its practicality awaits a detailed exploration of such worlds, an

inventory of their resources, a better understanding of the phenomenon of planetary habitability, and a future where the Solar System is opened to technological civilization as a new and expanding frontier.

Research in this multidisciplinary domain advances at a rapid pace and is very appealing to the public. But ethical considerations must also be taken into account. Even if we could terraform a planet, should we do it, and under what conditions (see Subsection 6.2.2 above)? Haynes (1990), among others, argued that terraforming raises new issues in ethics, so that

> we need from philosophers a new 'cosmocentric' ethics, and perhaps a revised theory of intrinsic worth ... [Such a] cosmocentric ethic would allow scope for human creativity in science and engineering throughout the Solar System.

Currently, we know too little about these far-away worlds in our Solar System, not to mention the exoplanets, and not enough about the Earth, to know whether life can really take root elsewhere.

## 6.6 HELLO TO OTHER LIFEFORMS?

Every civilization in history has wondered at some point or other whether alien life exists and if it could be technologically advanced enough to get in touch with us. Among the general public, as among scientists, some have looked forward eagerly and others with fear at the idea of such discoveries, which would have an extraordinary impact on our way of thinking and our lives. But supposing such alien civilizations exist and we do want to get in touch – how do we proceed to find them? And if we did find them, how would we treat them?

Sending messages out to space in the hope that they may be received by aliens who would then respond, or vice versa, has been a regular theme in science fiction. SETI, however, standing for the Search for Extraterrestrial Intelligence, is no such work of fiction. It is an organization dedicated to the search for radio signals from space that would mean that intelligent beings exist outside our planet, and it runs some of the most well-known scientific projects from Harvard University, the University of California, Berkeley. SETI projects use scientific methods to search for intelligent life on other planets. Amongst others, these searches include monitoring of electromagnetic radiation for any sign of transmissions from alien civilizations.

The United States government contributed to early SETI projects, but recent work has been primarily funded by private sources. In 1992, for instance, the US government granted funds for an operational long-term SETI programme (the NASA Microwave Observing Program, MOP), which aimed to conduct a large-scale survey of the sky but also to target 800 predefined nearby stars. MOP, however, was cancelled by the US Congress after a year. SETI supporters continued with private sources of funding, and in 1995 the non-profit SETI Institute of Mountain View in California resurrected the MOP programme under the name of Project Phoenix, which aims to study roughly 1000 nearby Sun-like stars.

The SETI Institute has in particular been collaborating with the Radio Astronomy Laboratory at UC Berkeley to develop a specialized radio telescope array for SETI studies, with the aim of coordinating numerous large antennas to search for radio extraterrestrial signals. The array concept in northern California, named the 'Allen Telescope Array' (ATA) after its benefactor Paul Allen, has a sensitivity equivalent to a single large dish more than 100 metres across. The first part of the array, 42 antennas, became operational in October 2007 (Figure 6.10), but in April 2011, the ATA as a whole was abandoned because of funding shortfalls. Operations were resumed on 5 December 2011, but whether the full 350-element array will see the light of day depends largely on available funding and technical results from the 42-element sub-array.

Until March 2004, when observations ceased, Phoenix observations were performed regularly from such telescopes as the 64-metre Parkes radio telescope in Australia, the 43-m radiotelescope of the National Radio Astronomy Observatory in Green Bank, West

FIGURE 6.10 The Arecibo radiotelescope. The image shows the bridge allowing access to the focal laboratory suspended above the telescope's centre. For colour version, see plates section. (Image courtesy of the NAIC – Arecibo Observatory, a facility of the NSF.)

FIGURE 6.11 The SETI Project: the Allen Telescope Array's first phase of 42 telescopes as completed. To build the full observatory (350 antennas), the SETI Institute will need additional funding. For colour version, see plates section. (Image © Zack Frank/Shutterstock.com)

Virginia, and the vast 300-m radiotelescope at the Arecibo Observatory in Puerto Rico (Figure 6.10). At that time, the project completed the largest ever search for signals in some 800 stars in the frequency range from 1200 to 3000 MHz. The search had quite some sensitivity at its disposal, and the negative result could also be interpreted as the Earth being in a 'quiet neighbourhood'.

The scope has now been expanded into the optical realm (optical SETI, or OSETI), which scans the sky for powerful light pulses from other stellar systems. Communication with extraterrestrial intelligence (CETI) is a branch of SETI focusing on composing messages that are transmitted to space and destined for alien civilizations, and also specializing in deciphering any messages that might be received in response. The research is based on all forms of language, from mathematics to pictorial, and on algorithmic communication and computational advances. One such communication plan was the well-known 1973 Arecibo message composed by Frank Drake and Carl Sagan (Figure 6.12).

According to some estimates given by those involved with SETI, roughly 2 million US dollars were invested in 2012 to keep the project going at the SETI Institute and approximately 10 times that to support all of SETI's activities around the world.

Contact with extraterrestrial civilizations is, however, not just a matter of deploying large antennas and space observatories: religious and ethical considerations may arise which are not to be neglected. Centuries of debate on the friendliness or malevolence of foreign civilizations have demonstrated that these questions are a sensitive subject for humanity. Although this experiment is extremely ambitious and complex, the potential reward is enormous. So far, no clear sign of extraterrestrial life has been found in any of the searches. But perhaps we should continue looking, or lend an ear to the sounds of space instead...

## 6.7  CONCLUSIONS FROM A PLANETOLOGIST'S POINT OF VIEW

As mentioned in the early parts of this chapter, space exploration of our Solar System and beyond, even with the prospect of scientific

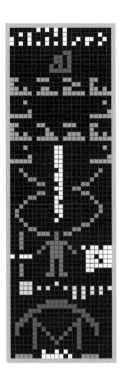

FIGURE 6.12 The Arecibo message as sent in 1974 from the Arecibo Observatory. From bottom to top: the Arecibo telescope; a picture of the Solar System; a human figure; the double DNA helix; the formulas of the sugars and bases in the nucleotides of DNA; the atomic numbers of hydrogen, carbon, nitrogen, oxygen and phosphorus, which make up deoxyribonucleic acid (DNA); and the numbers 1 to 10. For colour version, see plates. (Image by Arne Nordmann, courtesy of Wikimedia Commons.)

understanding and establishing an inventory of possible habitats, comes at a price.

So, first of all, let's look at the cost. The question of whether we should spend all that money in the face of the famine and poverty on our planet is not only legitimate, but pertinent and essential to answer. We may want to note that although most artistic, philosophical and scientific enterprises must have appeared useless in their early days (and remain so today for many people), for others they are essential because they are an integral part of humanity at the same level as engineering and medical applications and advances. How should we spend our resources, on our planet or beyond? Space exploration is expensive whether conducted by robots in our neighbourhood or by transporting people to far-away realms. But if we don't look beyond our local abode, we do not satisfy our natural curiosity, we do not learn about our world and our environment, and we may be less capable of protecting it. In any event, one thing seems important for a planetologist: that before risking lives by transporting people to explore other planets, we should always begin by sending appropriate robotic experiments to investigate the place, clear the land and open the way for more thorough and efficient exploration.

Second, we need to be extremely careful about protecting the environments. We explore at the risk of destroying not only so-far undetected ecosystems but also unique worlds that could teach us a lot about our own place in the Universe. Ethical questions about whether it is our right to colonize or transform exciting unique places in the Universe are also legitimate.

Lastly, we can always dream. We can investigate plans for possible future space stations and cities that might host humans; nothing can stop the technological and scientific advances that arise from our natural human curiosity. But we also need to keep as a top priority the saving of our own planet and ensuring fair and efficient use of its limited and diminishing resources.

# Further reading

RECOMMENDED LINKS

The NASA website at JPL with information on space exploration: http://jpl.nasa.gov
The European Space Agency (ESA) website on space exploration: http://sci.esa.int
Astrobiology Web: http://www.astrobiology.com
Astrobiology Magazine: http://www.astrobio.net
The planets: http://nineplanets.org
SETI Institute: http://www.seti.org/
The Extrasolar Planets Encyclopedia: http://www.obspm.fr/planets
The NASA Astrobiology Institute: http://nai.arc.nasa.gov
NASA/AMES Astrobiology: http://astrobiology.arc.nasa.gov
Centro de Astrobiologia, Spain: http://cab.inta-csic.es/
Planet Quest: http://planetquest.jpl.nasa.gov/
Space website: http://www.space.com
Talk origins: http://www.talkorigins.org
The International Society for the Study of the Origin of Life and Astrobiology
    Society: http://issol.org


REFERENCES

Allen, M., Cook, J.-R. & Weselby, C., NASA News topics 2010–190. (http://www.
    nasa.gov/topics/solarsystem/features/titan20100603.html)
Antoniadi, E., 1975, *The Planet Mars* (translated by Patrick Moore). Keith Reid
    Limited.
Bada, J. & McDonald, G., 1996, *Anal. Chem.* **68**, 668A–673A.
Bada, J. L., 2004, *Earth and Planetary Science Letter*, **226**, 1–15.
Badam, J. L., 2004, *Earth Planet Sci Lett.* **226**, 1–15.
Barton, N. H., Briggs, D. E. G., Eisen, J. A., Goldstein, D. B. & Patel, N. H., 2007,
    *Evolution*. Cold Spring Harbor Laboratory Press, pp. 93–95.
Bézard, B., de Bergh, C., Crisp, D. & Maillard, J.-P., 1990, *Nature* **345**, 508–511.
Blanc, M., Alibert, Y., André, N. *et al.* 2009, *Exper. Astron.* **23**, 849–892.

Boston, P., 2010, The search for extremophiles on Earth and beyond: What is extreme here may be just business-as-usual elsewhere. The Astrobiology Web, http://www.astrobiology.com/adastra/extremophiles.html

Brocks, J. J., Buick, R., Summons, R. E. & Logan, G. A., 2003, *Geochim. Cosmochim. Acta* **67**, 4321–4335.

Brown, T. M., Charbonneau, D., Gilliland, R. I., Noyes, R. W. & Burrows, A., 2001, *Astrophys. J.* **552**, 699–709.

Bruice, P., 2009, *Organic Chemistry* 6th Edition. Pearson.

Burrell, J. G., 2002, *Click4Biology.info*

Charbonneau, D., Brown, T. M., Latham, D. W. & Mayor, M., 2000, *Astrophys. J.*, **529**, L45–L48.

Chauvin, G., La Grange, A. M., Zuckerman, B. *et al.*, 2004, *Astron. Astrophys.* **425**, L29–L32.

Clark, R. N. *et al.*, 2010, *J. Geophys. Res.*, **115**, E10, cite 18 E10005.

Cooper, J., Cooper, P. D., Sittler, E. C. *et al.*, 2009, *Planet. Space Sci.* **57**, 1607–1620.

COSPAR, 2008, *Planetary Protection Policy* (revised 20 July 2008). COSPAR.

Coustenis, A. & Taylor, F. W., 2008, *Exploring an Earth-like World*. World Scientific Publishing.

Coustenis, A. & Taylor, F. W., 1999, *Titan: The Earthlike Moon*. World Scientific Publishing.

Coustenis, A., Atreya, S. K., Balint, T. *et al.*, 2009, *Exp. Astron.*, **23**, 893–946.

Coustenis, A. *et al.*, 2012, Life in the Saturnian system. In *Life on Earth and Other Planetary Bodies*. Springer.

Crovisier, J., Leech, K., Bockelée-Morvan, D. *et al.*, 1997, *Science* **275**, 1904–1907.

Dandouras, I., Garnier, P., Mitchell, D. G. *et al.*, 2009, *Phil. Trans. R. Soc. A* **367**, 743–752.

Encrenaz, T., 2008, *Ann. Rev. Astron. Astrophys.*, **46**, 57–87.

Encrenaz, T., Drossart, P., Feuchtgruber, H. *et al.*, 1999, *Planet. Space Sci.* **47**, 1225–1242.

Feulner, G., 2012, *Rev. Geophys.* **50**, RG2006.

Fitzpatrick, T., 2005, Calculations favor reducing atmosphere for early earth: Was Miller–Urey experiment correct? Washington University in St Louis Newsroom, http://news.wustl.edu/news/Pages/5513.aspx

Fogg, M. J., 1995, Terraforming Mars: a review of research, http://www.users.globalnet.co.uk/~mfogg/paper1.htm

Fogg, M. J., 1998, *Adv. Space Res.*, **22**, 415–420.

Gaeman, J., Hier-Majumder, S. & Roberts, J. H., 2012, *Icarus* **220**, 339–347.

Gonzalez, G., 1997, *Mon. Not. Roy. Astron. Soc.* **285**, 403–412.

Gowanlock, M. G., Patton, D. R. & McConnell, S. M., 2011, *Earth Planet. Astrophys.*, arXiv:1107.1286.

Hand, K. P., Chyba, C. F., Priscu, J. C., Carlson, R. W. & Nealson, K. H., 2009, Astrobiology and the potential for life on Europa. In *Europa* (R. T. Pappalardo, W. B. McKinnon & K. K. Khurana, eds.), University of Arizona Press, p. 589.

Hanel, R. A., Conrath, B. J. Jennings, D. E. & Samuelson, R. E., eds, 1990, *Exploration of the Solar System by Infrared Remote Sounding*, Cambridge University Press.

Haynes, R. H., 1990, Ecce ecopoiesis: playing God on Mars. In *Moral Expertise* (D. MacNiven, ed.), Routledge, p. 177.

Iess, L., Jacobsen, R., Ducci, M. *et al.*, 2012, *Science* **337**, 457–459.

Johnson, A. P., Cleaves, H. J., Dworkin, J. P. *et al.*, 2008, *Science* **322**, 404.

JUICE, 2011, Exploring the emergence of habitable worlds around gas giants. ESA/SRE(2011)18, December 2011 Assessment Study Report, p. 29.

Kasting, J. F., 1993, *Science* **259**, 920–926.

Lal, A. K., 2009, Searching for life on habitable planets and moons, http://arxiv.org/ftp/arxiv/papers/0912/0912.1040.pdf

Lammer, H., Bredehoft, J. H., Coustenis, A. *et al.*, 2009. *Astron. Astrophys. Rev.* **17**, 181–249.

LESIA http://www.lesia.obspm.fr/cosmicvision/tssm/tssm-public/

Lorenz, R., Lunine, J. & McKay, C., 1997, *Geophys Res Lett.* **24**, 2905–2908.

Lorenz, R. D. and the Cassini RADAR Team, 2008, *Planet Space Sci* **56**, 1132.

Mayor, M. & Queloz, D. 1995, *Nature* **378**, 355–359.

McKay, C. P. and Smith, H. D., 2005, *Icarus* **178**, 274–276.

Moses, J., Madhusudhan, N., Vissher, C. & Freedman, R. *et al.*, 2013, *Astrophys. J.* **763**, 25.

Muller, M., 2011, in *BIOS 100 Biology of Cells and Organisms*, Fall 2011, Hayden-McNeil.

Mumma, M. & Charnley, S., 2011, *Ann. Rev. Astron. Astrophys.* **49**, 721.

NASA *Protecting Life on Other Bodies*. Office of Planetary Protection guidelines.

*New Shorter Oxford English Dictionary*, 1993, Vol. 2.

Nic, M., Jirat, J. & Kosata, B., eds., 2006–. Chirality. *IUPAC Compendium of Chemical Terminology* (Online edition), DOI:10.1351/goldbook.C01058.

NRC, 2007, *The Limits of Organic Life in Planetary Systems*. Committee on the Origins and Evolution of Life of the National Research Council. National Academies Press.

Oro, J., 1961, *Nature* **191**, 1193–1194.

Owen, T., 1980, The search for early forms of life in other planetary systems – future possibilities afforded by spectroscopic techniques. In *Strategies for the Search for*

*Life in the Universe* (M. D. Papagiannis, ed.), Astrophysics and Space Science Library Vol. 83. Reidel, p. 177.

Raulin, F., Bénilan, Y., Coll, P. *et al.*, 2006, *Proc. SPIE* 6309, *Instruments, Methods, and Missions for Astrobiology* **IX**, 63090I, http://dx.doi.org/10.1117/12.675408.

Robinson, K. S., 2012, *2312* (science fiction). Orbit Books.

Sagan, C., 1973, *The Cosmic Connection.* Anchor Press/Doubleday, p. 47.

Schaeffer, L. & Fegley, B., 2007, *Icarus*, **184**, 462–483.

Shinnaka, Y., Kawakita, H., Kobayashi, H. *et al.* 2011, *Astrophys. J.* **729**, 81.

Smith, B. A. & Terrile, R. J., 1984, *Science*, **226**, 1421–1424.

Stevenson, D. 2000, Planetary interiors. In *Encyclopedia of Astronomy and Astrophysics* (P. Murdin, ed.), Institute of Physics Publishing, p. 1823.

Strobel, D., 2010, *Icarus*, **208**, 878–886.

Swain, M., Vasisht, G. & Tinetti, G., 2008, *Nature* **452**, 329.

Tinetti, G., Beaulieu, J. P., Henning, T. *et al.*, 2012, *Exper. Astron.* **34**, 311–353.

Turnbull, M. C., Traub, W. A., Jucks, K. W. *et al.*, 2006, *Astrophys. J.*, **644**, 551–559.

Udry, S., Bonfils, X., Delfosse, X. *et al.*, 2007, *Astron. Astrophys.*, **469**, 43.

Vogt, S. S. *et al.*, 2010, *Astrophys. J.*, **723**, 954–965.

Walsh, K. J., & Morbidelli, A., 2011, *Astron. Astrophys.*, **526**, A 126.

Ward, P., 2007, *Life As We Do Not Know It: The NASA Search for (and Synthesis of) Alien Life.* Penguin Books.

Wordsworth, R., 2012, arXiv:1106.1411v2

Wordsworth, R., Forget, F., Selsis, F. *et al.*, 2011, *Astrophys. J.*, arXiv:1105.1031v1 [astro-ph.EP].

Zubrin, R. M. & McKay, C. P., 1993, Technological requirements for terraforming Mars. AIAA, SAE, ASME and ASEE 29th Joint Propulsion Conference and Exhibit, Monterey, CA, 28–30 June 1993.

## RECOMMENDED BOOKS

R. T. Arrieta, *From the Atacama to Makalu: A Journey to Extreme Environments on Earth and Beyond* (Coqui Press, 1997).

Koki Horikoshi & William D. Grant (eds.), *Extremophiles: Microbial Life in Extreme Environments* (Wiley Series in Ecological and Applied Microbiology, 1998).

Michael Gross, *Life on the Edge: Amazing Creatures Thriving in Extreme Environments* (Basic Books, 2001).

David M. Karl (ed.), *The Microbiology of Deep-Sea Hydrothermal Vents* (CRC Series on Microbiology of Extreme and Unusual Environments, 1995).

Penny S. Amy & Dana L. Haldeman (eds.), *The Microbiology of the Terrestrial Deep Subsurface* (CRC Series on Microbiology of Extreme and Unusual Environments, 1997).

Cindy Lee Van Dover, *Deep-Ocean Journeys: Discovering New Life at the Bottom of the Sea* (Addison-Wesley, 1997). Cindy Lee Van Dover was the first female pilot of the research submarine *Alvin*.

Kim Stanley Robinson, *2312*, (Orbit Books, 2012).

C. Impey, J. Lunine & J. Funes (eds.) *Frontiers of Astrobiology* (Cambridge University Press, 2012).

J. Lunine, *Astrobiology: A Multidisciplinary Approach* (Addison Wesley, 2005).

D. A. Rothery, I. Gimour & M. A Sephton (eds.), *An Introduction to Astrobiology* (Cambridge University Press, 2003).

P. Davies, *Other Worlds* (Simon and Schuster, 1980).

F. Forget, F. Costard & P. Lognonné, *Planet Mars: Story of Another World* (Praxis Publishing, 2008).

F. Casoli & T. Encrenaz, *The New Worlds: Extrasolar Planets* (Praxis Publishing, 2007).

T. Encrenaz, *Searching for Water in the Universe* (Praxis Publishing, 2005).

T. Encrenaz, J.-P. Bibring, M. Blanc *et al.* *The Solar System* (Springer, 2004).

M. Ollivier, T. Encrenaz, F. Roques, F. Selsis & F. Casoli. *Planetary Systems: Detection, Formation, Habitability of Extrasolar Planets* (Springer, 2009).

M. Perryman, *The Exoplanet Handbook* (Cambridge University Press, 2011).

O. Botta, J. L. Bada, J. Gomez-Elvira *et al.*, *Strategies of Life Detection* (ISSI Space Sciences Series, Springer, 2008).

Vincent Coudé du Foresto, Dawn M. Gelino & Ignasi Ribas (eds.), *Pathways Towards Habitable Planets* (ASP Conference Series Vol. **430**, 2010).

M. Gargaud, B. Barbier, H. Martin & J. Reiss (eds.), *Lectures in Astrobiology*, Vol. I: *The Early Earth and Other Cosmic Habitats for Life* (Springer, 2006).

M. Gargaud, H. Martin & P. Clayes (eds.), *Lectures in Astrobiology*, Vol. II (Springer, 2007).

J.-P. Beaulieu, S. Dieters & G. Tinetti (eds.), *Molecules in the Atmospheres of Extrasolar Planets* (ASP Conference Series Vol. **450**, 2011).

G. Horneck & C. Baumstark-Khan (eds.), *Astrobiology: The Quest for the Conditions of Life* (Springer, 2002).

P. Cassen, T. Guillot & A. Quirrenbach (eds.), *Extrasolar Planets* (Saas Fee Advanced Course 31, Springer, 2006).

M. Gargaud *et al.*, *Encyclopedia of Astrobiology* (Springer, 2011).

*The Extrasolar Planets Encyclopedia.*

# Index